Pricing
for
Pollution

*An analysis of market pricing and government
regulation in environment consumption and policy*

WILFRED BECKERMAN
Balliol College, Oxford

IEA

Published by
THE INSTITUTE OF ECONOMIC AFFAIRS
Second Edition
1990

First published in December 1975

Second Edition January 1990

by

THE INSTITUTE OF ECONOMIC AFFAIRS

2 Lord North Street, Westminster,
London SW1P 3LB

Hobart Paper 66

ISSN 0073-2818
ISBN 0-255 36229-3

Printed in Great Britain by
GORON PRO-PRINT CO LTD
6 Marlborough Road, Churchill Industrial Estate, Lancing, W. Sussex
Text set in 'Monotype' Baskerville

CONTENTS

[5]

PREFACE TO THE FIRST EDITION

The *Hobart Papers* are intended to contribute authoritative, independent and lucid analyses to the understanding of the application of economic thinking to private and governmental activity. Their characteristic concern has been the optimum use of scarce resources and the extent to which it can be better achieved in markets using competitive pricing or by government using regulation based on centralised information and decision.

The subject of Hobart Paper 66 by Dr Wilfred Beckerman of Balliol College lends itself appropriately to this central theme. His subject is the extent to which economic 'rights' to the use of the environment is better disciplined by market prices in the form of charges or by direct control by government. The *Paper* reproduces (in Part II) the argument that Dr Beckerman outlined in a book published in 1974[1] and in the Minority Report (with Lord Zuckerman) to the 1972 Third Report of the Royal Commission on Environmental Pollution. He has also added Part I to explain the background of the argument for readers of *Hobart Papers*.

The 'consumption' of the environment can be analysed by economists in the same way as commodities and services in general. The environment—pure air, clean water and so on—is a scarce resource that is used in the production of goods and services by industry, public utilities, nationalised industries, local and central government. It must therefore be 'economised' so that it is used only to the point at which its social costs are covered by the social benefits. And this of course is equally true of scarce labour, equipment and capital used in production. The question is whether industry or government can be induced to economise its use more effectively by charges than by direct regulation.

The economic functions of a charge (a price) as a means both of reducing the use of a resource and also of imposing

---

[1] *In Defence of Economic Growth*, Jonathan Cape, revised and published in *Two Cheers for The Affluent Society*, St. Martin's Press, New York, 1975. We are indebted to both publishers for authority to reproduce the material in Part II in revised form.

[6]

a penalty on the amount used is not easy for non-economists to grasp. This may be perhaps why Dr Beckerman was unable to persuade more than one other member of the Royal Commission, Lord Zuckerman, to share the economic thinking embodied in the Minority Report. The majority of 7 members included 5 scientists (one in industry), a civil servant, and a cleric. Some were not opposed in principle to the use of charging as a means of controlling pollution but they thought it should not be introduced without further inquiries. In both parts of the *Hobart Paper* Dr Beckerman deals with the doubts and objections. What is surprising is that industry, which is presumably knowledgeable in the working of prices, does not seem to have understood his economic analysis.

The central value of the *Paper* is indeed the clear and cogent analysis of the economics of charges and direct regulation as alternative methods of controlling pollution. And of especial interest is Dr Beckerman's incisive dissection of seven objections to charging which indicates that they are, at least in part, founded in intellectual difficulty and error. The unanswerable argument used by Dr Beckerman to reply to the objections from industry is that people opposed to charges would not argue that their investment projects, or any other use of scarce resources, should be determined by direct state regulation.

Dr Beckerman's analysis is of especial interest in 1975 when the Layfield Committee appointed in 1974 to investigate local government financing has been gathering evidence from a wide range of sources and is expected to report by the end of the year. Although most of the evidence to it seems to have favoured revised or new forms of taxes, the Institute was asked to submit material on charging for local services. Dr Beckerman explains that local government would be very much involved in a system of charging for the use of the environment. There are now charges for the treatment of industrial effluent channelled to municipal sewers as a financial disincentive to pollute. Dr Beckerman's argument is that the principle should be applied more generally to discourage avoidable or uneconomic use of the environment.

A general objection to charging and the use of the price mechanism is that it bears more heavily on people with lower than on people with higher incomes. Here Dr Beckerman observes that the attempt to redistribute incomes lies at the root of most policies that deliberately misallocate resources. He cites examples from agriculture, tariffs, rents and the nationalised industries, in which controls are designed to even up incomes in favour of consumers or employees. And he concludes that the misallocation of resources might be avoided by scrapping such policies and redistributing income directly.

Moreover, on the use of charging in the control of pollution Dr Beckerman argues that the 'poverty' argument against charging is not even true, since the pollution is caused by industry and the charges would be borne by people with the higher incomes.

A fundamental general implication of Dr Beckerman's analysis is that it is wrong to regard the environment as an absolute that must be preserved at all costs. This again is where many non-economists have misunderstood the implications of economic analysis and drawn wrong conclusions. Dr Beckerman cogently demonstrates that it is appropriate to use the environment in the course of production if the loss of environment is exceeded by the gain in production of goods and services.

Dr Beckerman provides a convenient 'refute-it-yourself-master-key' for people confronted by objections to charging in principle (they apply to all scarce resources), and a secondary kit to refute objections on the ground of the impracticability of charging (the same objections apply to direct regulation).

The constitution of the Institute requires it to dissociate its Trustees, Directors and Advisers from the analysis and conclusions of the authors but it publishes this *Hobart Paper* both as an outstanding demonstration of economic analysis and of its applications to a difficult but main department of policy in which decisions must be made by government without much more delay. It will be found enlightening and stimulating by students and teachers of economics, by people in government who resist pricing, by conservationists, environmentalists and ecologists, and not least in industry where too often the functions of price are still misunderstood.

October 1975   ARTHUR SELDON

FOREWORD TO THE SECOND EDITION

Pricing for Pollution is fast approaching the status of a classic. When published in 1975 it set out in cogent and lucid terms the case for 'charging' for pollution as a means of environmental regulation. Dr Beckerman had then argued the case as a member of the Royal Commission on Environmental Pollution. His approach to the control of pollution was ignored, as was the problem itself. Governments prefer to rely on 'command-and-control' methods of regulation which fix standards and rely on pollution inspectorates to enforce the law backed by criminal sanctions. This approach is inherently inefficient, and because the penalties are infrequently imposed and often low, provides no real deterrent. Charges and other 'economic instruments' for environmental control seek to harness the pricing system to ensure that scarce environmental resources are protected and used efficiently. As Dr Beckerman argues in his original essay and new Introduction, the attractions of pollution charges and taxes are considerable, especially when compared with legalistic methods of control.

Yet market-like responses to pollution and other environmental problems are not exhausted by charges. Another approach is the development of property rights in the environment. These can be in the form of either the privatisation of un-owned or communal resources or the creation of a private property market in pollution. The first would simply seek to provide a legal framework and an initial assignment of property rights, leaving it to the market to decide how best the resources are used. If someone has an interest in husbanding elephants, eagles, the sea and atmosphere, then, like any other valuable commodity, they will ensure that it is optimally utilised and conserved. Poaching, the dumping of waste—using scarce resources as if they had no value—would simply not occur. But defining and enforcing property rights in the environment is often impossible, either for technical reasons or because the costs are simply much too high. In such cases other devices can be used. Tradeable pollution permits have been used in the United States. This effectively creates a market in pollution by awarding rights to firms to pollute up to a level which is thought accept-able, and then permitting them to trade in these rights. In this way the reduction in pollution is achieved in the least-cost way.

These approaches to pollution and environmental control provide regulators and governments with powerful weapons for protecting the environment without imposing excessive costs on industry, and ultimately the consumer. Dr Beckerman has drawn attention to charges as a solution in situations where it is difficult to define property rights in the environment. As governments respond to the pressures from environmentalists and concerned citizens they should pay as much attention to the techniques used to control pollution as to the case for more stringent environmental protection. Often the former is lost in the welter of arguments about impending doom and genuine concern. But ill-conceived intervention can sometimes be as bad as no intervention.

In the light of the increased concern over the environment the IEA has taken the opportunity to re-publish Dr Beckerman's classic analysis of charging for pollution. It is hoped that his analysis will reawaken in a new generation of students, scholars and policy-makers the timeless economic principles governing effective regulation. The original text skilfully scotches many of the fallacies which plague discussions of pricing solutions, and indeed the pricing system as a whole. The Institute takes much pleasure in this Second Edition, for it shows once again that one of our authors, who are asked to carry sound economic analysis of a problem to its logical conclusion without regard to short-term political possibilities, has found his views acceptable even if it has required 15 years for them to seep into the market-place of ideas. The Institute hopes by this timely publication to contribute to the growing debate over *how* best to protect the environment.

January 1990   CENTO VELJANOVSKI

AUTHOR'S INTRODUCTION
TO THE SECOND EDITION

The idea of replacing the usual direct controls on pollution by pollution charges has at least started to become respectable in Britain. For example, proposals along these lines in a recent report by Pearce, Markandya and Barbier received a very favourable reception in the British media and among interested groups.[1]

My original Hobart Paper No. 66 presented the full case for pollution charges, together with a detailed refutation of the many objections to such a policy that I had encountered in the course of advocating it back in the early 1970s. For the proposal to use pollution charges—that is, some form of taxation—to curb excessive pollution is, of course, not new. As explained in this *Hobart Paper*, I had urged the Royal Commission on Environmental Pollution, of which I had been a founder member from 1970 to 1973, to support pollution charges in its Third Report (1972), but of the eight other members of the Commission only one, Lord Zuckerman, shared my view, with the result that we were the only two signatories of the Minority Report to the Third Report of the Commission.[2]

In fact, the general proposition that taxes or subsidies, as the case may be, should be used to correct the misallocation of resources that usually arises when social costs do not equal private costs goes back at least as far as the great British economist, A. C. Pigou, who set out the relevant principles in his *Economics of Welfare* (1920). But my experience on the Royal Commission gave me an opportunity to work out in detail how this principle could be applied to the particular area of pollution control and how it compared with the methods of quantitative regulation and control that were then generally—and still are—in use.

[1] David Pearce, Anil Markandya and Edward Barbier, *Sustainable Development*, a report by the London Environment Economics Centre for the UK Department of the Environment, 1989, republished as *Blueprint for a Green Economy*, London: Earthscan, 1989.

[2] Royal Commission on Environmental Pollution, *Third Report: Pollution in Some British Estuaries and Coastal Waters*, Cmnd. 5054, London: HMSO, 1972.

The most striking lesson I learnt at the time was how wide was the range of fallacious objections to pollution charges advanced by those who seem to be psychologically averse to the use of the price mechanism in this particular activity. This included people who would never dream of replacing the price mechanism by quantitative controls in other economic activities. The objections came from all sides: fellow members of the Royal Commission, civil servants, scientists and, above all, the extremists who seemed to be directing environmentalist movements in those days and who condemned pollution charges as a 'licence to pollute', as if regulations specifying maximum amounts of pollution that firms could carry out without paying a fine were not also a 'licence to pollute'—and free at that! Consequently, my original *Hobart Paper*, which is reprinted here, contains a 'do-it-yourself-refutation kit' to deal with these objections.

Economic Case for Pollution Charges

On the positive side, pollution charges have several desirable economic attractions. First, they improve resource allocation at lowest cost. They ensure that pollution is reduced most in firms that can reduce it at least cost. And they provide firms with a continuing incentive to reduce pollution, in the same way that the pricing of labour or capital or raw materials provides firms with a continuing incentive to economise in their use.

But, secondly, since pollution charges are not designed to change the aggregate pressure of demand in the economy, they should not be allowed to affect the Budget balance. So the receipts from pollution charges should be offset by reductions in other taxes. And most other taxes—however justified they may be—generally distort the allocation of resources. For example, they distort the choice between work and leisure, or between investment in housing rather than in other types of saving, and so on. Hence, if some of the revenues from pollution charges are used to cut other taxes (or moderate the increase in taxes that would be required to carry out long-overdue environmental improvements), *there is a double gain.*

Thirdly, one of the obstacles to international agreements in this field is the tendency for the officials concerned to try to apply internationally the same type of uniform   quantitative controls they are accustomed to implement domestically. The economic principles underlying the pollution charge approach, however, clearly demonstrate the logical absurdity of uniform environmental standards. For the pollution charge approach should start from the principle that polluters should pay for the damage their pollution does to the environment. Since a given amount of pollution will do far less damage in some situations than in others, it makes no sense to impose uniform standards on everybody. But we shall never get other countries—or European Community officials—to accept this point if we do not even adopt the pollution charge approach at home.

Fourthly, another obstacle to international agreements is that poorer countries understandably feel that, when most of their citizens are not sure how they are going to get the next square meal, they cannot afford to make sacrifices now in order to improve the global environment for those of them who may survive to the year 2030 or whenever (at any rate long enough to enjoy it), let alone for the benefit of the rich countries who have been doing most of the polluting anyway. But if environmental protection is based on the correct economic principles, and not on quantitative controls and regulations, it should be easier to meet these legitimate fears in two ways.

Help for Poorer Countries via Pollution Charges

First, by demonstrating that (i) the resource cost of their pollution reduction would be lower if some price mechanism method is used rather than direct controls, and (ii) that since any extra revenue from pollution charges can be used to reduce other taxes, there is no reason, in principle, for the international competitiveness of poorer countries, on balance, to be threatened. Secondly, the above is not intellectual persuasion—it is fact. Pollution charges can be combined with concrete financial assistance to poorer countries that is specifically linked to environmental protection. For the rich nations could hand over a part of their revenues from pollution charges to the poorer nations in a way that is related to the latter's pollution abatement programmes.

For example, subject to certain reservations discussed in more detail in the main text of this *Hobart Paper*, a firm has much the same incentive to reduce pollution if it has to pay a tax on every unit of pollution it emits as it does if, instead, it receives a subsidy for every unit by which it reduces pollution. This is because the opportunity cost to the firm of an extra unit of pollution is the same in both cases. Rich countries could impose charges on their own polluters and hand over a part of the proceeds to poor countries to be used to subsidise their pollution abatement in this way. This would be equivalent to linking some aid specifically to pollution abatement. After all, the citizens of the rich countries expect to gain from global environmental protection as much as, if not more than, the poor countries. The fact that some of the pollution is produced in the poor countries is largely irrelevant to the equity question of who should pay to reduce it.

Of course, officials in the environment control business are accustomed to direct regulation and control, and the application of the charging principle to international agreement will require careful thought. But we do have time to think. So far the hysteria over global warming is based on evidence that is at least uncertain—if not quite as hysterical as was the 1960s scare that the world was on the brink of a new ice age. Of course, there is need for vigilance and for policies to protect the environment. Nor does there seem to be much doubt that human activity is likely to add significantly to atmospheric carbon dioxide levels. But a balanced view of the evidence suggests that there is no necessity for panic-stricken attempts to cobble together draconian international controls that are not only unnecessarily costly but that will also be unacceptable to most developing countries.

A Programme and Framework of Reform

Some forms of international pollution—such as pollution of the beaches—are sufficiently well understood for appropriate action to be initiated without much delay. Others, notably global warming, are far from clearly understood, and some governments, including our own, have refused to be stampeded into unwarranted, drastic controls. This breathing space should be put to good use, and would mean:

o starting to make serious progress in the application of pollution charges at home, rather than merely intensifying direct controls, which is all that seems to be advocated in, for example, the Environmental Protection Bill (December 1989);

o working out a framework in which pollution charges can be incorporated into international agreements; and

o exploiting the possibilities such a framework opens up for the richer countries to help the poorer in the interests of the whole human race.

[14]

Unfortunately, none of these negotiating advantages—and probably none of the advantages of a pollution charge approach to the problem domestically, let alone internationally—will be achieved if the policies continue to be negotiated by those who cannot understand novel proposals. Therefore the necessity for a full understanding of the superiority of charges over direct controls is now even more urgent than in the early 1970s. Having failed to convince the powers that be and most of my colleagues on the Royal Commission back in 1972 that their objections to the pollution charges proposals were unfounded, I went to special lengths in the *Paper* that follows to meet them all. Since I have little doubt that many of the same objections will be raised again. I can only hope that this time my analysis finds a more receptive audience.

Grounds for Optimism

I have some grounds for this hope, for there is little doubt that the climate of opinion is more favourable now to some form of pollution charges than in the early 1970s. There are several reasons for this shift of opinion.

First, there has been a general and widespread change of mood over almost all the world against planning and quantitative controls. Of course, this is partly a rejection of the political authoritarianism and repression associated with such economic régimes. But it is also partly, if not largely, because of the failure of widespread economic *dirigisme* to produce the goods. Benevolent and liberal-minded at heart as President Gorbachev may be, the transformation that he has initiated owes much to the recognition that the economic failure of detailed centralised planning was turning much of the Soviet bloc into impoverished Third World countries.

Secondly, as I predicted in my *Hobart Paper* in 1975, one of the many weaknesses of reliance on regulations was that the degree of enforcement depended on fashion, which changed from time to time. It could shift about from pollution to nuclear war, to crime and violence on the street, to drugs, to AIDS, to homelessness, and so on. Taxes, however, tend to go on and on, and once the machinery to collect a tax is set up the inspectors carry on doing the job year in year out, whatever current fashion may be most influential among local government or central government authorities or legislators concerned with retaining electoral support. As a result, the environment has tended to be neglected for most of the last 15 years. After a sudden spurt of interest in the early 1970s the world became distracted by the oil shocks, and then inflation, and then unemployment, and only recently has the environment come back into the limelight again. Many environmentalists now recognise the necessity for a pollution charge system to be put in place before the world forgets about the environment again for a few more years.

Thirdly, one of the reasons why the environment has come back into fashion is that, rightly or wrongly, there is now more of a basis for serious concern about the environment. This is partly the result of continued governmental neglect, for the reasons I have set out. But it is also partly because of the concern with threats to the globe—notably the ozone layer and global warming. Even if some of these alarming predictions turn out to be greatly exaggerated they do at least appear to be better founded than, say, the Club of Rome's infamous *Limits to Growth*. As a result, we have seen a growth of environmental protection pressure groups that seem to be more professional in their analysis and less prone to wild exaggeration than was the case 20 years ago. They are consequently taken more seriously.

Furthermore, as I pointed out long ago in my 1974 book, *In Defence of Economic Growth*, as people get richer their priorities change. Very poor people worry about getting a square meal the next day and a roof over their heads. When their basic needs are satisfied they begin to worry about other ingredients of their welfare, including the environment. During the last two decades people in advanced countries have become richer. Hence one would expect that they should become more concerned with the environment.

Fourthly, in my early 1970s proposals I defended pollution charges on cost grounds. I argued that pollution charges must reduce pollution more cheaply than regulation. My arguments were based partly on theory and partly on the few empirical studies that had then been made and that confirmed the theoretical prediction. But many more studies have since been made and various OECD publications document the much wider body of knowledge that is now available concerning the cost-effectiveness of the policy of pollution charges.[1] Consequently, those who are seriously concerned with improving the

[1] See, for example, the summary in Jean-Philippe Barde, 'The Economic Approach to the Environment'. in *The OECD Observer*, June/July 1989, which also contains a useful survey of pollution charges in various other OECD countries.

environment, rather than with merely making dramatic gestures, are gradually learning that if, from the point of view of society as a whole, pollution charges are a cheaper way of reducing pollution, there will be more pollution reduction with a charging mechanism than with a regulation mechanism.

Fifthly, over the years some form of price mechanism to combat pollution has been introduced more widely, so that it is more difficult to assert now that charges are impracticable. In May 1989, for example, President George Bush announced a programme to introduce pollution rights, with an accompanying market in such rights, as a measure to combat air pollution. In fact, many OECD countries have introduced measures of this character. In Sweden, where the use of pollution charges is most advanced, there is even a Parliamentary Committee on Environmental Charges, whose function is to make proposals concerning the level of charges appropriate in specific instances.

First Priority: Experiment with Pollution Charges in the UK

Of course, as indicated above, much pollution nowadays has such a global character that international action is required. In spite of declarations of support for the 'polluter-must-pay' principle,[1] progress towards satisfactory international action has been very slow. But the practical problems of reaching satisfactory international agreement are enormous. It must, therefore, seem tempting to simplify matters by aiming at simple quantitative targets. Since such targets would, however, mean that pollution reduction would impose a far heavier economic burden, particularly on poorer countries, the long-run chances of making adequate progress along these lines are slim. Until it has been tried it will not be known how far we can go with the much cheaper pollution charge approach. And the place to start trying is at home, as some other countries have already done—successfully. And this implies that those concerned with environmental protection should begin by trying to understand the basic principles. It is hoped that this *Hobart Paper* may make some modest contribution to that objective.

January 1990

WILFRED BECKERMAN
Balliol College, Oxford

---

[1] As, for example, in the declaration adopted at Ministerial level by the OECD Environmental Committee on 20 June 1985, and 10 years earlier in the OECD's *The Polluter Pays Principle*, Paris: OECD, 1975.

THE AUTHOR

WILFRED BECKERMAN was born in 1925 and educated at Ealing County School and Trinity College, Cambridge. He was a Lecturer in Economics, University of Nottingham, 1950-52; worked for the OEEC/OECD in Paris, 1952-61; National Institute of Economic and Social Research, 1962-63 (Member of the NIESR Executive Committee since 1973); Fellow of Balliol College, Oxford, 1964-69; Economic Adviser to the Board of Trade, 1967-69; Professor of Political Economy, University of London, and Head of the Department of Political Economy at University College, London, 1969-75, from which he resigned in order to return to his old post as Fellow and Tutor in Economics at Balliol College in Autumn 1975. He was a Member of the Royal Commission on Environmental Pollution, 1970-73.

Dr Beckerman's publications include: *The British Economy in 1975* (with associates) (1965); *International Comparisons of Real Incomes* (1966); *An Introduction to National Income Analysis* (1968); (ed. and contributor) *The Labour Government's Economic Record* (1972); *In Defence of Economic Growth* (1974); *Measures of Leisure, Equality and Welfare* (1978); (with associates) *Poverty and the Impact of Income Maintenance Programmes* (1979); (ed. and contributor) *Slow Growth in Britain* (1979); (with Stephen Clark) *Poverty and Social Security in Britain since 1961* (1981); (ed. and contributor) *Wage Rigidity and Unemployment* (1986).

PART I

Pricing:
Objections and Refutations

I. ECONOMICS AND RATIONALITY

Anyone who would like to write a thesis on the way that irrationality, emotion and prejudice dominate the discussion and implementation of economic policy could do no better than to study pollution policy. That personal interest, class alliance, and political passion, rather than simple economic logic, should influence the debate over, say, prices and incomes policy, or redistributive taxation, or nationalisation of industry, and so on, is hardly surprising. They are all major issues that have a profound influence on the distribution of incomes and that are also related to basic objectives concerning the way society should be organised.

Objectives and methods

But neither of these considerations applies to the pollution policy proposed in this *Paper*. This is essentially that the best way to keep pollution as near as possible to socially desirable amounts is by some form of pollution charge or tax on the amount by which polluters 'pollute'—i.e. use up—the environment. In other words, what is proposed is a price mechanism solution rather than a direct control method. Most people would probably rapidly agree on the objectives of pollution policy, so that differences of opinion on objectives are unlikely to explain the opposition to the pricing policy proposed here. Nor does the choice of the price mechanism, in preference to direct control, imply a systematic bias in the effect of pricing on income distribution in the economy as a whole. Whichever method is adopted, some resources have to be used in order to reduce pollution, and there is no economic reason to distinguish between the two methods with respect to which class of the population, in general, will have to make the corresponding sacrifice.

Nor is it possible to attribute to the opponents of pollution charges a general ideological preference for the direct controls, production quotas and norms, and so on that are associated with Soviet-type economics.[1] Opponents of pollution charges

---

[1] However, one reviewer of my book *In Defence of Economic Growth* objected to my 'neo-classical' approach, the term 'neo-classical' being one of the general smear-words characteristic of a certain school of economic thought today, and that, in the eyes of some of the less gifted followers of this school, damn the arguments of those accused of being 'neo-classical' without there being any need to go into the logic of the matter or to produce positive scientific propositions that one can discuss rather than merely match with counter-adjectives and abuse.

include the business community, who would vigorously oppose direct controls over the amounts of labour or raw materials and so on they use but who seem to prefer direct controls over the amounts of the environment they 'use up' when their pollution reduces the clean air or clean water available to the rest of the community. Pollution charges also tend to be opposed by most civil servants concerned with pollution policy, although they, too, would be alarmed at the suggestion that they are thereby, perhaps, revealing a hitherto concealed and subconscious predilection for Soviet-type economic regulation.

Ignorance and over-reaction

Given the absence of the usual obvious reasons why various groups should oppose pollution charges, it can only be concluded that they must do so out of ignorance of the argument, so that it is worth trying once more to draw it to their attention. It is to be hoped that the slightly revised extract (Part II) from *In Defence of Economic Growth*,[1] plus the additional material (Part I) will demonstrate clearly why the arguments advanced by an astonishing alliance between some left-wing economists, the business community and bureaucrats against some form of pollution charge or tax are simply wrong. Meanwhile, whereas almost all taxes we do have tend to lead to a misallocation of resources in the economy, one of the few forms of tax that, by contrast, would tend to improve the allocation of resources is one of the only taxes we do *not* have.

There is, however, one group of people whose opposition to pollution charges is rational in the sense of being consistent with their general views on related topics: those who are hostile to economic growth in principle. In the first place, much of this opposition to economic growth reflects an emotional over-reaction to some of the obvious disamenities of modern life. Such a state of mind is not one in which cool, logical analysis of policy measures to deal with one of the disamenities of the modern world is likely to flourish. Secondly, although there is, of course, an element of truth in many of the fashionable criticisms of modern economies, the unwillingness of the anti-growth school of thought to take the good with the bad and to recognise the enormous benefits that economic growth

---

[1] Jonathan Cape, London, 1974. (The USA edition, under the title *Two Cheers for the Affluent Society*, was published by St. Martin's Press, New York, 1975.)

has conferred on humanity in general (if not the middle classes in particular—at least in some respects) reflects partly a deep emotional rejection of the uglier features of industrial, urban civilisation. This psychological frame of mind is, of course, one in which *all* pollution is believed to be intolerable, even though mankind has polluted his environment in one way or another throughout the ages.

Consequently, the starting point of the economic analysis of pollution presented in Part II is unacceptable to the passionate anti-growth cohorts, since this starting point is that society should be prepared to accept some 'optimum' pollution rather than deprive itself of the goods and services that would have to be sacrificed if pollution were to be brought below this optimum. And the extreme anti-growth cohorts cannot accept the notion that *some* amount of pollution is socially desirable, since it then becomes immediately obvious that it is necessary to know the criteria by which this optimum is determined. And this, in turn, leads to the difficulty that, since they do not know these criteria, they are unable to demonstrate that the current level of pollution is too high; for perhaps it is too low. If we cannot define optimum pollution, we cannot specify the reference point by which the present amount of pollution can be judged too high or too low.

Thirdly, even those opponents of economic growth who would accept that there is a (non-zero) optimal amount of pollution still dislike the price mechanism solution to the pollution problem, since the inevitability of excessive pollution has for long been part of the anti-growth case. Hence, the suggestion that there is not really any serious pollution problem, since socially optimal amounts of pollution could easily be achieved by a pollution charge, destroys one major plank in the anti-growth platform. Furthermore, the basic economic analysis of the pollution problem—more than the analysis of other parts of the anti-growth case, such as the alleged inevitable exhaustion of raw materials—makes it very clear that the problem is essentially that of resource allocation in a given period. And the distinction between optimal allocation of resources at a point of time and optimal allocation of resources over time—which is what the growth problem is really about, or ought to be about—also exposes a fatal flaw in the anti-growth case. For the elementary analysis of the pollution problem clearly demonstrates the nature of a misallocation of resources at a given point of time; and once this proposition is clearly grasped it becomes immediately obvious that such misallocations do not depend on economic growth for their existence. It becomes quite clear that stopping economic growth tomorrow would not stop factory chimneys from emitting too much smoke or sulphur dioxide, nor automobiles from emitting too many exhaust fumes, nor even oblige all local authorities to instal adequate sewerage systems.

Indeed economic growth[1] has been, and is likely to continue to be, the major means by which society will be induced to *reduce* pollution to socially optimal levels. As nations grow richer they become more willing to devote resources to improving the environment. Whether they will take advantage of the increased scope for doing so is another matter and depends on how wisely, at any point of time, they allocate resources between the claims of the environment and other claims such as for housing, health, education, public infrastructure, private consumption, and investment. Part II is designed to indicate how such a correct allocation of resources can be pursued.

II. THE PRICE MECHANISM
AND THE 'SOCIAL OPTIMUM'

The 'optimum' and 'optimal use of resources'

Up to this point I have been using the terms 'optimum' and 'optimal use of resources' without indicating what they are supposed to mean. This probably has not detracted from the general drift of the argument so far, but it has become necessary before going further to clarify the notion of optimal use of resources. If an economy is fully employed and its labour and capital stock are used as intensively as is desired given society's conventions on hours of work and so on, it will be necessary to take resources from one use in order to devote them to another. The resources will be optimally allocated between all uses when society would lose more welfare by taking a unit of resource away from one use than it would gain by putting it to some other.

[1] As I argue at some length in Chapter 5 of the book.

If, for example, a marginal unit of resources were taken away from the production of, say, homes, society would lose some welfare as a result of the reduced output. If this unit were used to increase output of, say, clothes, it would add to welfare. Resources are allocated optimally if, at the margin, any switch of this kind would mean a loss of homes that would exceed society's valuation of the additional (marginal) output of clothes.

Particularly in practical application, the criteria of optimal allocation are, of course, not as simple as that. Innumerable complex conceptual and practical problems surround the measurement of the effects of switching resources from one use to another. And one major kind of limitation on the above definition of optimum allocation of output is that it takes no account of 'distributional' considerations: the effect of different ways of allocating resources on the general equality of income distribution or on the fortunes of particular groups.[1]

Although distributional considerations are certainly important and, furthermore, the distribution of incomes will affect the relative prices and costs that enter into the micro-economic analysis of resource allocation, it is also important to know what optimal resource allocation would be *if* distributional considerations could be left aside. This is not simply because, in many concrete cases, they may not be important or relevant. It is also because few people would argue that income distribution is the *only* thing that matters: that it does not matter if we all starve as long as everybody is in the same boat. Most people, therefore, would accept some 'trade-off' between improvements in income distribution and a reduction in total output and income available for the community. But, again, it is impossible to know what this reduction in output would have to be if we cannot say how output could be maximised by an optimal allocation of resources. This is the point generally overlooked by the smears against 'neo-classical' economics.

The 'hidden hand' and the social optimum

Hence it is important to know what the optimal allocation of resources means and why, under certain conditions, it will not

---

[1] These are two quite distinct distributional effects. We might accept a certain amount of misallocation of resources if this is believed to be the price that has to be paid to achieve more equality of income. But we might also have views on the relative importance of raising the welfare of different groups irrespective of the effect on income distribution.

be automatically obtained by the free-market mechanism. This topic has been the subject of a long debate that goes back at least as far as Adam Smith's 'hidden hand', according to which the pursuit by everybody of his own economic advantage would usually tend to produce the optimum social advantage. The basic idea can be outlined briefly.

The value to society of an additional unit of output of a product may be taken as its price. If the output of all goods (and services) were pushed to the point where the costs to society ('social costs') of producing the 'last' (marginal) unit were just equal to its price, resources would be optimally distributed and the pattern of output would be optimal. Assuming, for simplicity, that an increase in output would lead to a rise in its social costs, an increase in output of a product beyond the point where the costs of the marginal unit equalled the price would mean that the additional units of output were produced at a cost to society that exceeded its value to society as measured by its price.

It would be desirable if all firms produced their goods up to the point where the social costs of the last unit of output equalled the price of the product, since it would mean that the socially optimal pattern of output was being produced. In practice, under competitive conditions, firms will have an incentive to push production of any good up to the point where the costs *to them* of producing the last marginal unit will be equal to its price. For it is at this level of output that they will maximise their profits. In order, therefore, for the behaviour of firms to yield the socially desirable pattern of output all that is required is that the costs of production to firms should be the same as the costs of production to society as a whole. For in that case, insofar as firms try to maximise profits by pushing output to the point where price equals their own marginal costs, they would also be pushing output to the point where price equals social marginal costs, which, as we have seen, is also the point of optimal social output.

Qualifications to the 'hidden hand'

There are, however, five qualifications even to a much more detailed and rigorous statement of the 'hidden hand' doctrine. In brief, there is a wide measure of agreement among economists that the 'hidden hand' cannot be expected to achieve the social optimum in certain situations. First, competition is rarely perfect. Secondly, as indicated above, different income distributions will affect prices and costs. Thirdly, there are some types of goods, known as 'public goods', such as the classic examples of lighthouses, defence, and so on, from which it is virtually impossible to exclude users who do not wish to pay for them once they are provided for those who do. Hence, it is virtually impossible to supply 'public goods' at a socially optimal level through reliance on the free-market mechanism. Fourthly, there are 'merit' goods, such as a certain quantum of education or health or law and order, which many people would regard as a basic inalienable right that should be available equally to everybody irrespective of income. In other words, some people may not object to a rich man being able to buy a bigger car or have a bigger house than a poor man but they might think it immoral that differences in wealth should also permit the rich to pay for better education or health, or preferential treatment from the police or the fire brigade. Hence, even where it would be technically possible to operate a private market for the goods and services in question—unlike lighthouses or defence, where it is not—society might decide, as a matter of basic value-judgement, that it prefers them to be distributed according to 'need' rather than ability to pay.

'Public goods' and 'merit goods' would, therefore, be better provided through the public sector. This means that, to preserve the desired pressure of demand in the economy, the government will have to raise some taxes. In general these taxes will tend to distort the allocation of resources, though there is no reason in theory why this distortion should be severe and little evidence, in practice, that it is so.

Externalities, 'spillovers' and pollution

There is a fifth reason for supposing that the free market will not always lead to the socially optimal pattern of resource allocation, so that government intervention is required to achieve the social optimum. There is a class of economic activity characterised by simple 'spillover' effects, or 'externalities', of which pollution is an obvious example. In these activities, optimum output does not result from the market mechanism because some of the costs of the activity are not borne by those responsible for it. The classic example is the factory chimney which imposes costs on neighbours from its smoke and sulphur dioxide and so on.[1] These costs will not normally enter into the calculations (nor the accounts) of the factory. Hence, if the factory produces up to the point where *its own* costs are just covered, at the margin, by the value of its output (as measured by its price), it will have pushed output too far from the social point of view. At this margin, total social costs—which will include both its own costs and the pollution costs borne by others—must exceed the value of its marginal output.

Role of a pollution tax

In other words, output is not optimal because of a divergence between social costs, which must include the pollution, and private costs. This divergence violates a key assumption of the 'hidden hand' doctrine, which is logically necessary (though not sufficient by any means) for the subsequent conclusion that the output of competing producers is socially optimal. But if, in these cases, the reason for the failure of the market to produce socially optimal output is a divergence between private and social costs, the remedy is simple; impose a tax (or a subsidy where private costs exceed social costs) to bridge the gap. In the pollution example all that is required is a tax on the firm equal to the value of the pollution damage done by additional output. This would bring the firm's private costs including the tax into equality with the social costs. If then the firm pushes output to the point where, at the margin, its private costs *including the tax* equal the price of the product it will also ensure that social costs are equal, at the margin, to the value of the output, which is what is required for an optimal level of output.

This simple example shows basically why a pollution tax or charge happens to be a rare example of a tax that *improves* the allocation of resources rather than worsens it. This is why it is extraordinary that the authorities refuse to use it, although

---

[1] Pollution is not produced solely by firms: consumers also 'pollute' the environment, as when they leave litter lying about, or use noisy lawn-mowers, and so on. Also, pollution arising in the course of productive activities is not confined to the private sector guilty of a desire to maximise profits irrespective of the effects on the environment. Many major polluting activities take place in the public sector, as in public transport (noise as well as air pollution), public utilities such as electricity production, publicly-owned iron and steel plants, or even the 6 a.m. refuse collection of some public authorities which may disturb the sleep of many householders.

they use (if largely unavoidably, in order to pay for mainly legitimate public expenditures) a proliferation of taxes that, taken by themselves, distort resource allocation.

There are, no doubt, all sorts of reasons why various groups object to action that appears to make elementary economic sense. The highly emotion-charged reaction of most of the anti-growth school of thought to any suggestion that pollution is simply a matter of misallocation of resources at a moment of time has nothing to do with the growth debate. Other groups who are opposed to pollution charges are no doubt motivated by different considerations. In the absence of any opportunity to subject them all to psycho-analysis it is not possible for me to speculate on their inner motivations. Whatever they are, the arguments advanced against pollution charges and in favour of some direct controls over pollution are always couched in terms which have the superficial appearance of appeal to rationality, but it requires very little thought and understanding of the problem to see that these objections are all misguided. In Part II many of the customary objections to pollution charges are examined and their fallacies exposed in detail.

III. PRICING POLLUTION: THE MASTER KEY

Scarcity the key to understanding

The essence of the issue can be put in the form of the following 'refute-it-yourself master-key' to the arguments advanced against pollution charges. Pollution is objectionable because it constitutes the 'using up' of a resource to which we attach a value, such as clean air, or water, or peace and quiet. When a firm (or a neighbour or the local authority) emits smoke or dirty effluent into the environment it is depriving us of some of the clean air or water that we enjoy. If we could always provide ourselves with *unlimited* amounts of clean air or water we would not mind how much of it was 'used up' by the polluter. But we become worried about it when the environment is a *scarce* resource so that, for instance, the pollution of a river or a beach means that we are deprived of the possibilities of enjoying its amenities. The key point is that the environment is a scarce resource and pollution is, in effect, a use of this resource.

Once this proposition has been grasped the rest is easy. Faced with any argument against using the price mechanism (by pollution charges) to control pollution and thus allocate the use of the environment amongst the many who want to use it, all that is required is to ask *why that argument does not apply to the use of the price mechanism to control all other uses of scarce resources?* Some of my colleagues on the Royal Commission on Environmental Pollution who refused to accept the pollution charges proposed here,[1] objected to them on the ground that large firms would be able to pay the charge and hence pollute on a large scale whereas small firms would not be able to do so. Armed with the refute-it-yourself-master-key, however, the alert reader no longer needs prompting from me to ask why, in that event, they do not object to the use of the price mechanism to allocate, say, labour or raw materials, which are scarce resources like the environment, on the ground that this merely enables large firms to buy much more labour and raw materials than small firms.

A second (reserve) key

The reader may also find it useful to learn how to use a second, reserve, key. This is to be used when the argument is not so much why, in principle, charges should not be used to control pollution but in terms of the practical difficulties of applying this principle. It must be conceded that there are immense practical difficulties in the way of deciding what is the ideal optimum charge that should be imposed in individual cases and in measuring the amount of pollution caused by a polluter. These practical difficulties are often proffered as if they constituted valid reasons for preferring direct controls to charges. To deal with this sort of argument the universal-refute-it-yourself key is to ask why the same difficulty does not

[1] The Third Report of the Royal Commission on Environmental Pollution (Cmnd. 5054, HMSO, 1972), which was concerned with pollution of rivers and estuaries, was accompanied by a Minority report, signed only by Lord Zuckerman and me, in which we advocated pollution charges to deal with the pollution. The other seven members of the Commission were either opposed to the whole idea in principle or were unwilling to propose the use of charges without receiving further advice from the civil servants. Neither Lord Zuckerman nor I had any objections, of course, to hearing the views of civil servants, even though it would be more customary—or at least more appropriate—for civil servants to take advice from Professors of Economics on economic issues of this kind rather than for Professors of Economics to take advice from civil servants. Furthermore, this happens to be an issue on which almost all economists are in agreement.

apply to the use of the direct controls? Some of my colleagues on the Royal Commission, and most of the scientists in the government services concerned with pollution, maintained that it was not possible to measure precisely any regular flow of, say, effluent into a river, so that it would be inequitable or inadequate to charge according to the inaccurate measurement taken at a particular point in time—say, every Monday afternoon. They would argue, for example, that if a family were charged for its domestic gas consumption on the basis of a measure taken at one point in the week (say, Sunday morning), either the gas consumer would claim that it exaggerated his normal weekly consumption, or, conversely, the charge might be much too low if, for example, the family ate out on Sundays.

All this is perfectly true, *but the alternative*—a direct control which presumably has to be backed up by a fine for failure to respect it—*is open to the same objection.* If a household is given a maximum permitted weekly gas consumption (like pollution) and if, in addition, the extent to which it has respected this limit is based on a reading of the flow of gas consumption at noon on Sunday when the Sunday joint is being cooked, we should be equally outraged at such a violation of natural justice.

The same sort of reply applies to the argument that it would be useless to settle on an agreed time at which the discharge of pollution is to be measured to calculate the charge. Here it is argued that polluters will naturally arrange to time the flow of their pollution so that it is low at the time it is measured for charging purposes. But the same objection applies if it is to be measured for checking whether a quantitative control is being respected.

Measurement essential

Thus the position adopted by some of the participants in the debate about pollution charges, namely that they might become a more practical proposition only when continuous monitoring of pollution becomes available, completely misses the point. Measurement is required just as much to ensure that direct controls are respected as to ensure that pollution charges are levied. Controls without measurement are farcical. The introduction of more accurate and continuous monitoring instruments merely means that *both* methods—charges or direct controls—can be better operated; it does not change their *relative* merits.

For the same reason we need not be impressed by the argument that direct controls can often take the relatively simple form of agreeing with the firms on the pollution abatement technique they should adopt—such as the type and size of water purification plant or height of chimney, and so on—and then checking that they install it. For the effect of the equipment on the subsequent degree of pollution cannot be assessed by hunch or theoretical calculations (which are usually impossible because of the complexity of the variables that enter into the determination of pollution, such as local water and air conditions). Some measurement, however rough and ready, of the impact of the equipment is essential if pollution control is to be taken seriously. But once this is done the measurement can be used as a basis for a charge; it is just as good for charging as for assessing whether the equipment satisfies the objectives of the direct controls. The cosy relationship that probably exists between authorities responsible for the control of pollution and the polluters is, of course, easier to maintain if all that the authorities have to do is check that the local factory has installed the agreed equipment. But pollution control policy should not be operated solely in the interests of enabling some officials to maintain a cosy relationship with polluters.

Incentive to reduce pollution

In Britain these officials have done a very good job, and a good relationship between them and polluters is no doubt desirable and valuable as a means of implementing anti-pollution policy. But too high a price can sometimes be paid in the interests of preserving cosy relationships, including refusal to implement a method of control that has the advantages claimed for pollution charges (Part II). These advantages include not merely the improvement in the allocation of resources at a point of time given known techniques, but also a permanent and continuous incentive to firms to *reduce* their pollution charges by finding economic methods of further reducing pollution, in exactly the same way that they have a continuous incentive to find ways of economising their use of any other input (labour, land, machinery, etc.) into their productive process for which they have to pay. Is it too much to hope that a careful reading of Part II will convert at least those groups that want to take pollution seriously but that have so far seemed content to make dramatic gestures about it?

IV. SUMMARY AND CONCLUSIONS

1. Since charges for pollution neither profoundly affect the distribution of incomes more than direct government regulation, nor concern the basic objectives of social organisation, opposition to them must be based on irrational prejudice rather than on economic reasoning.

2. Whether the price mechanism or direct government control is used, resources have to be consumed to regulate pollution, and there is no economic reason to suppose the sacrifice has to be borne by some particular income groups more than others.

3. People in industry and in the Civil Service are inconsistent in so far as they prefer direct controls to pollution charges as a means of regulating the 'use' of the environment, although they would not support direct government regulation for other scarce resources, such as labour or raw materials.

4. The opposition to pollution charges is derived essentially from lack of understanding of the argument for charging. Basically, almost all existing taxes tend to misallocate resources; a pollution charge (tax) would tend to improve the allocation.

5. Opposition to charges also comes from those who oppose economic growth in principle, and who believe all pollution is intolerable and do not understand that it is economic for society to 'use' up the environment up to a point, i.e. where, at the margin, the costs to society of doing so are offset by the extra goods and services produced. The aim should be not 'no pollution' but 'optimum pollution'.

6. Excessive pollution is simply a misallocation of resources at a moment of time. Misallocation of resources is not the outcome of economic growth. Ending growth would not stop pollution. And growth is itself the major means by which pollution can be checked.

7. The use of scarce resources is optimised when more welfare would be lost by removing a unit of resources from one use than would be gained by putting it to another.

[33]

8. A major limitation on this criterion of optimal allocation is the effect on the distribution of incomes, although these effects may be secondary and they must be weighed against the effects on total output, i.e. there is an optimum 'trade-off' between more equality in income and a reduction in total income.

9. The optimum allocation of resources is not yielded automatically by the free-market mechanism because competition may be imperfect, prices and costs reflect possibly undesirable income distributions, the socially desirable supply of 'public goods' (which have to be provided free) would not be forthcoming, nor of 'merit goods', and some activities (like pollution) have external effects that require taxes or subsidies to produce the optimum amount.

10. The anxiety about the despoliation of the environment arises from its *scarcity*. If clean water, pure air, space to stand and stare, etc., were unlimited there would be no objection to pollution, for example, of one river or beach, because there would be no sacrifice. This proposition provides the 'refute-it-yourself-master-key' to the answer to arguments against charges. The objection that charges would enable large firms to pollute on a larger scale than small firms is not made against the use of prices for labour or raw materials on the ground that large firms can buy more than small firms. Yet all are scarce resources used in production.

11. The further objection that charges are administratively impracticable can be refuted by a second (reserve) key: that precisely the same difficulties apply to the control of pollution by direct government regulation because the same measurement and checking of pollution is required for both methods.

12. The argument that direct control can often be simpler than charging, because it can be arranged by agreement between officials and polluters on anti-pollution equipment, is unimpressive. The price of such cosy relationships can be too high in sacrificing the advantages of charging: improvement in the allocation of resources at a given point in time, and the incentive to firms to evolve industrial processes that reduce pollution.

PART II

Comparative Economics of Pricing and Regulation

I. POLLUTION AND THE FIRM

Pollution creates a 'problem' because it is essentially an 'external diseconomy'. Its harmful effects—i.e. the costs it imposes on the community—do not enter into the calculations of the producer responsible largely because either

(i) the environment polluted, such as the air or the water, is not clearly anybody's 'property', so that the polluter does not have to pay the owner for using it up, or

(ii) the environment is somebody's property, as with many common-law rights to clean air or water, but these property rights are generally difficult or expensive to protect, and payment cannot usually be extracted from polluters.

In either case, polluters will not have the same incentive to 'economise' in their pollution as they have in things they have to pay for, such as labour, capital or raw materials. The *economic* problem, therefore, is to find the best way of inducing polluters to economise in pollution up to the point where a further reduction would cost society more than the resulting benefits, i.e. where the marginal cost exceeds the marginal benefit.

Property rights and pollution

To understand more fully why the absence of property rights in the environment leads to a 'pollution problem', and the pros and cons of alternative remedies, it would be desirable to set out the full economic analysis. But most readers probably do not wish to embark on a crash course in these concepts for the purposes of better appreciating the subsequent steps in the argument. Consequently we shall try to present the main features of the analysis in a fairly rough and concise manner.

There are two main ways in which one can examine the role of pollution in the productive activity of the firm. Polluting consists of producing a pollutant, such as smoke, which adulterates part of the environment, such as clean air. Hence, pollution can be analysed either in terms of *the pollutant*—the smoke or the effluent or the noise—produced as a by-product of the activity, or of *the clean environment* destroyed, or 'used up', by the pollutant. The second approach regards the clean environment used up like any other input into a firm's productive process: labour, or capital, or raw materials. When a firm

[37]

pollutes the environment it is using up clean air or pure water. The first approach, perhaps more conventional in economic theory, consists of regarding the pollutant as an undesirable by-product of the productive activity. Thus, for example, air pollution associated with steel production can be regarded either in terms of the smoke produced as a by-product of the steel or of the clean air 'used up' to produce steel, together with the iron, labour, fuel, and so on.

If pollution is regarded as undesirable, it differs from ordinary products because it carries no price. Normal products carry prices. These prices are related to the benefits they confer on the purchasers, and they also provide an incentive to produce the products as long as the returns producers get exceed the costs. Since pollution is harmful, it is obvious that, by analogy with the positive prices for ordinary products, pollution should carry a negative price. This would both correspond to the *negative* benefit it confers on people and would also constitute the required *disincentive* to producers to supply this undesirable product. In other words, we would have the exact counterpart of the prices for 'goods', which correspond to the benefits to the consumers and which provide incentives for their production. One is simply the converse, or the mirror image, of the other; in the same way that ordinary products are regarded as 'goods', pollution is a 'bad'.

A negative price?

What is meant by a negative price? A positive price means that the more the firms sells the more it receives. A negative price should mean that the more the firm sells the more it pays. A tax on a product is one form of negative price, since the more of the product that is 'sold' the more tax has to be paid. In so far as a pollutant does not carry a negative price there is

(i) no relation between its price and the damage it confers on people, and

(ii) no disincentive to produce this undesirable by-product.

Hence, an excessive amount of it is certain to be produced.

The same conclusion is reached if pollution is analysed from the second point of view: as a free input of the environment into the productive process. The polluter is seen as having no incentive to economise in his use of the free factor of production, the environment (or 'the facility to pollute'). If a steel producer finds a way of making steel that economises in fuel

[38]

or labour or capital equipment but produces more smoke (i.e. more input of clean air), he will tend to adopt it since he saves money by using less of the normal inputs and it costs him no more to increase his use of the environment. Similarly, if the firm finds it cheaper to switch to the distribution of milk in plastic containers instead of in glass bottles, he will do so whether or not the production and disposal of plastic containers imposes higher pollution or other external costs than do glass bottles. It is true that technological innovation in some industries has reduced the amount of pollution per unit of output; for example, the switch from coal to other sources of fuel over the last decade in Britain, which has been partly the result of technological progress, has helped to reduce air pollution in British cities. But this benefit to the environment has been fortuitous, and one cannot rely on technological change to reduce pollution in the absence of the appropriate incentives.

It should not be thought that it is only private industry that, in its concern for profits, tends to push pollution beyond the socially optimal point. Much pollution is produced by public authorities for the very same reason as for firms, namely the 'externality' character of pollution. Many sewage works in Britain are inadequate, partly because the benefits from better installations—cleaner effluent and hence cleaner rivers— would be enjoyed by communities further downstream. Hence, the benefits are often 'external' to the local authority that would have to pay for the improved sewage works. That is why, in general, 'there are no votes in sewage'. One of the motives for the recently proposed re-organisation of river and water services in Britain into much larger 'Regional Water Authorities' is to remedy this weakness of the system.

Nor is it true that pollution is unique to capitalist countries on account of their subservience to the profit motive.[1] For many years now the authorities of the Soviet bloc have been increasingly concerned with very serious pollution, and have been obliged to take increasingly strict measures to fight it.[2]

Resource costs of reducing pollution
If a polluter is induced to reduce pollution, he will incur costs in doing so. He will usually have to replace some of his use of

[1] For example, assertions to this effect in Howard Sherman, *Radical Political Economy*, Basic Books, New York, London, 1972, pp. 74-7.

[2] Discussed in Chapter 2 of *In Defence of Economic Growth*, pp. 44-6.

the environment by making more use of other factors of production, such as labour or capital (for example, a taller chimney or a water-purification plant). In effect he will be obliged to adopt a different technique of production from the one he would choose when the environment was free. Since these other factors of production carry a price (for example, the wages of labour), his costs of production must rise when he uses more of them to cut pollution. The more he reduces pollution, the more it will cost him. Furthermore, in general, the more pollution is reduced the more difficult technically, and hence the more expensive, it becomes to reduce pollution by *a further unit*. In economic language, the more pollution is abated the higher, in general, are the 'marginal' costs of abatement—the costs of reducing pollution by a further unit.

The costs that matter to society are the real resources the polluter has to use to economise on his hitherto free use of the environment (or to cut his output of the undesirable pollutant by-product). For these resources can no longer be used by society for other purposes.[1]

In other words, one way or another, firms reduce pollution by substituting other factors of production for it. (It is because of this *substitutability* aspect of the reduction in pollution that it is sometimes easier to analyse the economics of pollution as an input into the productive process rather than as an undesirable by-product.) Since, to make less use of the environment, firms have to make more use of other factors of production, the latter are no longer available to society for other purposes, such as the production of food, clothing, consumer durables, machines, motor-cars, medical supplies, houses, and so on. Thus these costs of pollution abatement are also costs to society, and the more pollution is cut the more society must sacrifice food, houses, etc. that it could have obtained with the resources.

The problem, therefore, is not how to prevent *all* pollution but *how far* society should sacrifice other goods in order to *reduce* pollution. To solve this economic problem we have to examine the benefits from pollution abatement.

[1] The economist's term is 'opportunity costs'.

II. THE BENEFITS OF POLLUTION ABATEMENT

The benefits of reducing pollution simply equal the damage the pollution had been doing. The evidence on the costs of reducing pollution and the benefits to be derived therefrom (i.e. the damage done by pollution) is reviewed in the book.[1] Here we are concerned with the principles.

The damage done by pollution may take many forms: damage to health, loss of amenity, industrial change (such as corrosion of metals or deterioration of dyes), damage to crops or wild-life, and so on. It is by no means confined to economic damage to industrial or other output. The economist includes any loss of human welfare resulting from pollution in the social damage from pollution abatement. Nor need this damage be confined to the present or the immediate future; the economist would include the benefits to be reaped in the long term.

Like the costs of pollution abatement, the benefits of abatement are not likely to vary in exact proportion to the amount by which pollution is abated. At very high levels of pollution a given reduction in pollution may bring substantial benefits, but when pollution has been reduced to relatively low levels the gain to welfare from further reductions may be relatively very small. A small amount of polluted effluent in a river may merely make the water look less than perfectly pure and clear. A slight reduction in pollution from such low levels will probably make a negligible difference to welfare, so it would be uneconomic to spend large sums on it. At the other extreme, if pollution is very high, a small reduction may make a large difference to the level of dissolved oxygen in the water and hence determine whether fish can easily survive in it and, generally, whether it has an acceptable appearance and smell. It would then be worth spending much more to cut pollution by a relatively small amount.

'Optimum' pollution

In general, therefore, the more pollution is cut,

(a) the more it will cost society to reduce pollution by a further unit, and

(b) the less will society gain from cutting out a further unit of pollution.

[1] *In Defence of Economic Growth*, Ch. 7.

And, clearly, it makes no sense for society to push pollution abatement beyond the point where the costs are larger than the benefits. 'Optimum' pollution is the amount at which this point has been reached, i.e., at which the social costs of reducing pollution by a further unit just equal the social benefits of doing so, and where a further ('marginal') reduction in pollution would cost more than the value of the further ('marginal') benefits. [1]

In defining optimum pollution in this way the economist is not expressing a value-judgement about what is included in the costs and the benefits. Anyone is free to define them as he likes. The only value-judgement implicit in the definition of optimum pollution, therefore, is that society should maximise its welfare. Unless one dissents from that proposition it is difficult to dissent from the definition. In particular, it is not true that the definition of optimum pollution necessarily leads to an undesirably large amount of pollution on the ground that many of the costs of pollution are 'external'. The definition does not exclude these costs, but there is no reason to believe that special weight should be attached to them. External costs are no more costly to society, pound for pound, than any other costs, including the costs of abatement (which merely reflect, if indirectly, the value that society puts on the goods and services that could be produced with the resources involved). It is merely an accident of the legal and institutional organisation of society that some costs are 'external' to the person responsible.

Under the proposed re-organisation of the water services in Britain, for example, the sewage authorities and the river authorities will be combined in much larger administrative units, so that the costs of pollution caused by inadequate sewage, which had hitherto been external to the sewage authorities, will now become internal to the new larger organisations. But this administrative re-organisation cannot change the costs of the pollution to society, or the costs of sewage. Similarly, that airline operators do not have to pay to soundproof the homes of people who live near airports is the result of particular legal and institutional arrangements and does not mean that the costs to society of soundproofing would be different. Nor would the costs of a given amount of sound-

[1] The final part of this definition assumes that abatement costs rise, and benefits fall, as pollution is successively reduced.

proofing change if the law were amended so that the airlines had to bear them. These costs would no longer be 'external' to them and would enter into their internal calculations of how much noise their aircraft should make. Such a change in the institutional arrangements might induce them to spend more money on quieter engines and hence reduce soundproofing needs. But this is irrelevant to the resource cost to society of a given piece of soundproofing—say, double-glazing a window in a certain manner: it does not depend on *who* is legally responsible for bearing the cost.

Qualifications to the criterion

There are, however, many qualifications that can be made concerning the application of the optimisation criterion. (These qualifications are an integral part of the body of economic theory, not omissions from it.)

First, in practice it is very difficult to measure the costs and benefits of pollution. These difficulties lie first of all in our fundamental ignorance of the technical scientific relationships between changes in physical amounts of pollution and physical variables such as the incidence of bronchitis, the speed of metal corrosion, the loss of fish life, and so on. They are not primarily caused by the inability of economists to attach monetary values to these factors, though even if the precise physical data were provided by the scientists and technologists it would be very difficult to attach price-tags to them. Some of the implications of these difficulties for policy will be considered below (pp. 62-65), but they have no bearing on the principle of trying to maximise social welfare by equating the marginal costs of pollution abatement with the marginal benefits.

Secondly, the significance attached to estimates of the costs and benefits of pollution abatement is not entirely free from value-judgements. For the relevant prices and costs that enter into the estimates reflect the existing social and institutional arrangements of society. The manner in which income is distributed as between labour and capital, or the socially conventional view on the appropriate rate of profit, will affect the relative prices of goods embodying different proportions of labour and capital. Relative costs and prices will also be affected by the degree of monopoly tolerated in society. The effect of this sort of consideration on the costs of pollution can be illustrated by an extreme example. Suppose that immigrants into Britain were allowed to work only in laundries. The price of laundry services would then be much lower than if the laundry industry were in the hands of a tight white Anglo-Saxon monopoly. The valuation of the damage done by smoke from a factory chimney to, say, a workman's shirts would be much less, since it would not cost him so much to have them laundered.

Hence, the damage done by pollution would be valued *by the market* more cheaply than if the price of laundry services were kept higher under a different legal and institutional framework.

Optimum pollution and income distribution

Another, and related, qualification to the optimum criterion concerns income distribution. As well as reducing *total* social welfare (in the absence of corrective measures), pollution also affects its *distribution*. In so far as the costs of pollution are not borne by those who cause it, or by the purchasers of their products, but by people who happen to be victims, some of the total welfare of society is being redistributed away from the victims in favour of the other groups. Manufacturers (or their shareholders) who can pollute free of charge make bigger profits than if they were obliged to cover the full social costs of their production, including the external costs generated by their pollution. And the purchasers of their goods and services buy them at lower prices than if the prices covered the full social costs. Hence, the manufacturers and consumers of the products gain at the expense of the victims of the pollution.

Such redistributive effects may not always make the distribution of economic welfare less equal than it would have been otherwise. Sometimes the polluters will be relatively poor people and the victims may be richer: where a poor community cannot afford better sewage, the rich owners of yachts moored downstream suffer from the untreated effluent flowing into the river. Because of the effect of a change in resource allocation on the distribution of income or welfare, the effects of a policy to treat pollution should be taken into account as long as income distribution is accepted as part of social welfare.

It is always open to somebody to make the value-judgement that he is simply not interested in income distribution. A more acceptable reason for ignoring the distributional consequences of pollution policy would be that if the incomes are thought to be distributed too unequally, this defect should be dealt with anyway by measures designed to redistribute them *directly*, rather than *indirectly* by measures designed to deal with resource allocation. For, it can be argued, if resources are allocated 'optimally', so that total economic welfare is maximised, it will always be possible to distribute this welfare so that some people are better off without anybody being worse off than if resources are not allocated optimally. In simple terms, with a bigger cake it must be possible for some people to have bigger slices without anybody necessarily having a smaller slice. Hence, it would be argued, it is best to allocate resources optimally, since this course will produce the biggest possible cake, and then, if we do not like the way it is shared out, we can always change it by redistributive taxes and subsidies, taking care that these themselves do not misallocate resources.[1]

Misallocation caused by redistribution

There is much force in this argument, and it is probably true that income-distribution considerations are at the root of most of the policies that deliberately misallocate resources in the economy. For example, agricultural prices are often supported at artificially high levels in order to distribute income in favour of farmers; tariffs are levied on imports to protect those engaged in the home production of the goods; rents are controlled to distribute income in favour of the occupants; nationalised industries are subsidised to distribute income in favour of the consumers or the employees; and so on. One wonders how much better off everybody could be if all these resource-misallocation measures—almost all defended, in the end, on income-distribution grounds—were scrapped, and other measures adopted to redistribute income directly.

On the other hand, the limitations on the optimisation criterion (pp. 43-44), or the assumptions that would have to be made about the degree of resource misallocation in the rest of the economy, are considerable. Furthermore, it is not at all easy to devise direct methods that have no adverse effect on

[1] It is for this reason that economists tend to favour 'lump sum' taxes and subsidies, i.e. taxes and subsidies that are not proportional to any variable, such as the amount of one's income or the prices of goods, since it is the latter that tend to distort resource allocation one way or another.

resource allocation, and that, in the end, have as much effect on income distribution as expected. By contrast, a simple tax or subsidy on some product may have a relatively clear and pronounced effect on income distribution compared with its possible adverse effect on resource allocation.

Welfare economics and the price mechanism

These are complicated questions, and they have been the subject of much sophisticated discussion in the literature of what is known as 'welfare' economics. Most economists would probably agree that the main contribution they can make is to remind the policy-maker of all the relevant effects of any policy, and to attempt to rank these effects—such as the damage done by pollution, the way it affects different groups, and the way measures to reduce it would affect these groups. Most economists would also probably accept that

'If he succeeds in this task it will almost certainly become more widely appreciated that tinkering with the price mechanism is one of the more feasible and generally satisfactory ways of securing whatever distribution of wealth is desired.'[2]

In saying all this, therefore, we are not saying that the economic principles for optimum pollution are irrelevant. On the contrary, we are saying that they are not all that simple, and that the relevant economic theory has by no means neglected many valid reservations that have to be made to any simple rule. This is very different from abandoning any attempt at a rational policy for pollution, which usually leads to the adoption of some absurd rule of thumb, such as that is not only grossly over-simplified but that bears no relation to any starting principle of what it is we are trying to optimise. The optimisation criterion given above is, at least, derived from the objective of maximising total welfare.

The reasons why this rule may not always achieve this objective are well known in economic theory, as also are the adjustments that have to be made, in principle, in order that the rule remains consistent with the original objective.

It is surely preferable to operate in this context of maximising social welfare than in one which does not even seek to promote

[1] J. de V. Graaff, *Theoretical Welfare Economics*, Cambridge University Press, London, 1967, p. 161.

[46]

the highest possible social welfare. In fact, when alternative rules are suggested it usually turns out that the rule proposed is simply designed—if quite unconsciously—to promote the special interests of a small group in the community at the expense of the community as a whole. This *may* be socially desirable and acceptable, for one of the functions of society is to protect legitimate minority interests. But such policies should reflect the conscious and deliberate decisions of society, and should emerge from a rational analysis of the alternatives, not from an obscurantist rejection of any attempt to derive criteria, in a logical manner, from generally agreed objectives.

III. PRICE INCENTIVES FOR POLLUTION ABATEMENT

We have seen that the optimum amount of pollution is where the social costs of a further (marginal) unit of pollution abatement equal the social benefits. We have also seen that firms will not normally have any incentive to abate pollution at all, so that pollution will be pushed well beyond the optimum point. What is required is that the firm or the producer should have the same sort of incentive to economise on the use of the environment as he has to economise on other inputs into his productive process. One such incentive would be to make the producer bear the costs of his pollution. In this way his costs of production would reflect the true full social costs of his productive activity.

A pollution tax

Subject to certain assumptions, a producer will have an incentive to reduce pollution to the socially optimum amount provided he pays a pollution charge or tax that equals the cost to society of the pollution—i.e. a charge equal to the damage done to society at the point of optimum pollution. For firms will reduce pollution up to the point where the costs of their doing so are less than the tax they would otherwise pay. They will not reduce pollution further because it would cost them more to do so than to pay the tax. If, therefore, the tax is set to equal the (marginal) social benefits of abatement at the optimum level of pollution, firms will reduce pollution up to the point where their costs of doing so are also equal to these (marginal) social benefits of pollution abatement. And this is the socially optimum amount of pollution: the costs of a further unit of abatement are equal to the benefits. This merely corresponds to the more common-sense general proposition that, subject again to some well-known reservations, if producers have to pay the social cost of an input (in this case the clean environment) they will tend to use it only up to the socially optimum amount.

So far the sort of incentive to firms to reduce pollution that we have discussed has consisted of some kind of tax (or charge) per unit of pollution that they create. But, in practice, to provide the producer with the incentive to economise in his use of the environment, it does not matter much whether it consists of a tax that he has to pay for every unit of additional pollution he causes or a subsidy for every unit by which he reduces his pollution. The incentive to him to reduce pollution is exactly the same in both cases. A tax is merely the most obvious form of incentive to a producer to economise in the use of the input concerned. It also bears most resemblance to the payment that he would have to make if the environment were 'owned' by, say, private individuals and he had to settle some price with them in order to compensate for his use of their clean air or clean water. But the subsidy method would produce the same results as long as it is a subsidy *per unit* reduction in the firm's pollution, in exactly the same way as the pollution charge should be a charge *per unit* of pollution that the firm still causes.

For what matters, in principle, in deciding how to allocate resources are what the economist calls 'opportunity costs': the costs of what the firm gives up, or sacrifices, by allocating resources in one way rather than another or by making one decision rather than another. And the opportunity cost to a firm of creating an additional unit of pollution is either the tax he pays on it or the subsidy he foregoes as a result of not eliminating it.

Tax or subsidy?

Taxes and subsidies are not equivalent. The subsidy method is undesirable in the long term. It is a satisfactory substitute for a pollution charge only if it is related to the amount by which pollution is reduced below some initial level. If it were to be paid indiscriminately to firms for not polluting the environment it would be open to potential polluters to work a sort of 'protection racket', i.e. to set up business in a highly polluting activity in order to claim the subsidy for shutting down their polluting activity. It would then become rather like the story of the American farmers who are paid not to produce crops and who decide to increase their income by increasing the amount of the crops they will not produce! To prevent this situation, which could result in pollution rising instead of falling, it is essential that the abatement subsidy be related to the *initial* levels of pollution, and that it should not be applied to the avoidance of *new* pollution. But since the case for using the subsidy rather than the charge method would be on equity and income-distribution grounds—i.e. to reduce pollution without inflicting a loss of income on the polluters—the subsidy should, in general, be applied only to existing pollution.[1]

The subsidy method is feared to lead to more pollution than the tax method since it reduces the costs of the polluting activity instead of raising them, as with the pollution tax. At first sight it may certainly seem that the subsidy does reduce the costs of the polluting activity (steel-making might be an example). But this is not true as long as it is a subsidy per unit by which pollution is reduced and not, say, a subsidy on pollution-removal machinery. For in so far as there is the same price incentive to reduce pollution, firms will have to use more of other inputs such as labour and/or machinery in place of the environment, in exactly the same way as if the incentive had been provided by a pollution tax. Since these other inputs have prices, the firms' outlay on them must rise irrespective of the reason why the firm is induced to use more of them. Furthermore, the cost of the environment input (i.e. pollution) to the firm is exactly the same with either method—the tax or the subsidy foregone—so that the total costs of production, comprising the costs of the conventional inputs (labour, capital, etc.) plus the cost to the firm of its use of the environment, have risen in exactly the same way whichever method is used.

Direct compensation?

Indeed, pollution charges or abatement subsidies (generally known by the pejorative term 'bribes') are by no means the

---

[1] A. P. Lerner, 'Priorities and Efficiency', *American Economic Review*, September 1971. As Lerner points out, there are no grounds in equity for using the subsidy method for abating new pollution.

only two forms of 'tinkering' with the price mechanism that will tend to provide the same incentive to the firm to reduce pollution to the optimum amount. Other forms include a system by which the victims of the pollution pay the polluters according to the value to them of the reduction in pollution, or by which the polluters are obliged to pay the victims according to the damage done by their pollution. Both these methods, subject to assumptions, should lead to exactly the same result in the incentive they give to polluters to reduce their pollution. Whether the cost to them of an additional unit of pollution takes the form of a tax on it, or of the subsidy foregone, or of the payment they must make in compensation to the victims, or of the payment they forego from the victims for each unit by which they fail to reduce pollution, it creates the same *opportunity cost* of pollution, and hence should have the same effect on the degree to which pollution is reduced.

Effects on income distribution

But all these methods have different effects on income distribution. For example, with the tax method the proceeds should, in principle, be used to reduce other taxes (or increase government expenditures), for otherwise the pressure of demand in the economy would fall. This method, therefore, tends to benefit taxpayers (or the recipients of the other government expenditures) at the expense of the producers and consumers of the products concerned.

Thus the income-distribution effects of any mechanism must always be examined (pp. 44-46). Imagine an economy in which slavery was tolerated and widely used. To ensure the optimum allocation of resources in the economy, the optimum use of slave labour would be that at which the product of an extra unit of the labour just equalled the extent to which the slaves disliked it. This optimum could be achieved by making slave-owners pay a tax per hour of slave labour, like the tax on pollution. Slave-owners would then economise in their use of it and would reduce their use of slave labour to the socially optimum point where the addition made to output by a further unit of labour just equalled the tax. From the resource-allocation point of view this would be fine. But the slaves might not like it. They might prefer a price mechanism in which the price to be paid for their labour was not a tax paid to the state but a wage paid to them. Resource allocation would be the same (subject to the usual static assumptions), but the distribution of income would probably be very different indeed.

'Pollution rights'

Yet another method of providing an incentive to arrive at the socially optimum amount of pollution is the system of 'pollution rights' expounded by Professor J. H. Dales,[1] which has attractions. This is the system under which the authorities decide what they think is the desirable level of pollution (as they have to do under *any* system) and then issue, on the market, 'rights' to this amount of pollution, allowing the equilibrium price to be settled on the market. Firms that need to pollute as part of their productive process will bid for these rights. Possession of a pollution 'right' entitles a firm to carry out the specified amount of pollution, which is the same as limiting its obligation to reduce pollution. Firms that can reduce pollution cheaply will not want to buy as many rights as firms that would find it very costly to do so. Market imperfections apart, the price at which the rights will settle will be the same as the optimum tax (or subsidy).

One of the advantages of the pollution-rights method, as Professor A. P. Lerner has argued,[2] is that it is a way of using the price mechanism in a situation where uncertainty about the amount of damage done by the pollutant is combined with a fear that it could rise sharply if the optimum amount were exceeded. In such cases,[3] the authorities want to feel assured that pollution does not exceed some fixed quantitative upper limit. They may not wish to run the risk of over-estimating the extent to which a pollution charge would induce firms to abate pollution, or they may want to guard against the likelihood that conditions could change rapidly or that pollution would fluctuate around the optimum if the charge were kept constant. In such cases the authorities could simply operate on the amount of pollution they thought optimal, rather than its price, and issue pollution rights to this amount. They might miscalculate the costs of abatement to firms, so that the price at which the permitted quantity rights settle would differ from

---

[1] *Pollution, Property and Prices*, University of Toronto Press, Toronto, 1968.

[2] A. P. Lerner, *op. cit.*

[3] Examples include any product that becomes highly toxic when its concentration exceeds certain limits by *small* amounts, e.g. cyanide.

the one which they would have adopted if they had used the tax method. But they would still be safe in the sense that pollution would not exceed the desired amount, and the allocative advantages of the price mechanism would still apply.

IV. THE ADVANTAGES OF POLLUTION CHARGES

Broadly, a pricing system offers four advantages in pollution control.

(i) *'Allocative advantage'*

First, the 'allocative advantage' of pollution charges is essentially that, if all firms are subject to a uniform charge per unit of damage done by their pollution, those that can reduce pollution most cheaply will do so more than those that face relatively high or steeply rising costs. In other words, more of any given amount of pollution abatement will be made by firms that can do it most cheaply; they will use least resources for this purpose and hence least deprive society of these resources for other purposes, i.e. minimise opportunity costs. It is like using the price mechanism to ensure that shirts are produced as economically as possible, since the price mechanism will tend to ensure that most shirts will be produced by the firms that can produce them most economically. Such empirical evidence as is available also confirms that the charges system is cheaper than that of direct controls.[1]

The imposition of direct controls on pollution corresponds to the use of production quotas or 'norms' according to which firms are given production targets in the form of direct quantitative regulations. This method is generally unlikely to ensure that goods are produced by the firms best able to do so and by the most economical methods. With certain exceptions, it is not

[1] The pioneer work, in this connection, is the famous Delaware study, the economics of which have been described by Edwin Johnson in 'A Study in the Economics of Water Quality Management', *Water Resources Research*, Vol. 3, No. 2, 1967. See also Allen Kneese and Blair Bower, *Managing Water Quality: Economics, Technology, Institutions*, Johns Hopkins Press for Resources for the Future, Baltimore, 1968, and A. Kneese in J. Rothenberg and Ian Heggie (eds.), *The Management of Water Quality and the Environment*, London: Macmillan, 1974, in which a similar analysis of the Potomac is described. A similar conclusion emerges from Ernst and Ernst (consultants), *A Cost Effectiveness Study of Air Pollution Abatement in the Greater Kansas City Area*, US Dept. of Health, Education and Welfare, Washington DC, 1969.

the type of economic-policy instrument used in Western countries. The use of direct regulations for the control of pollution amounts precisely to such a system of production quotas and norms.

The same forces which tend to make the price mechanism a cheaper means of producing most goods apply to the abatement of pollution. If a uniform pollution charge is imposed at a stretch of river or estuary, all polluters will tend to abate their pollution up to the point where it would cost them more to abate further than the charge they pay per unit of pollution.[1] In other words, at the margin, the cost of pollution abatement is equal in all firms, since it is equal to the charge made to all firms. Contrast this with the use of some direct control, such as a regulation that all firms must reduce their pollution by a uniform percentage, or to some uniform amount. This will obviously involve very *high* marginal costs of pollution abatement for some firms and *low* marginal costs for others. Clearly, the same total amount of pollution abatement can be obtained if some of the abatement is switched from the firms where it is costly to those where it is cheap; and savings of this kind can be made by switching up to the point where the marginal costs of further abatement are equal for all firms. This is the situation to which the pollution-charge system tends to lead.

(ii) *Economising incentive*

The second advantage is that the individual firm will also have an incentive, under a price-mechanism scheme such as a pollution charge, to find the cheapest way to reduce its pollution, whereas many kinds of direct regulation take the form of laying down precise instructions on the steps that firms must take to reduce their pollution, such as raising the height of their chimneys or changing to an alternative productive

---

[1] In principle, a uniform charge does not mean that it does not vary according to the damage done by different physical units of pollution. A uniform charge means uniform in terms of the charge per unit of damage done. For example, a uniform charge per unit of pollutant applied to different firms along a stretch of river, which took no account of the major variations in the damage done by a given amount of that pollutant according to their precise location on the river, would be sub-optimal. Failure to allow that a 'uniform' charge is uniform in this sense, not in terms of unadjusted physical units (though, in practice, this will often be the best proxy variable), appears to be responsible for the arguments deployed by J. L. Stein in 'Micro-Economic Aspects of Public Policy', *American Economic Review*, September 1971.

technique. With a pollution charge, some firms will find it cheaper to change their raw-material input, others may carry out more re-cycling, others may institute more effluent-treatment plant inside the firm, others may change location, and so on.

Furthermore, firms will have a continuing incentive to experiment and to seek new and more economic methods of reducing pollution, for the more they do so the more they save on pollution charges, in the same way as the more they find ways of reducing labour per unit of output the more they save on wages.[1] By contrast, if they are given directions to reduce pollution by a certain amount, and possibly also by specified means, they have little further incentive to reduce pollution even more than the regulations require. Whatever the system of direct control, firms have little or no incentive to do better than the control limit given to them. For this is invariably a maximum pollution limit which they must not exceed; but they win no prizes for falling short of it. With the tax system, however, the more they reduce pollution the less tax they pay. It is not enough to say that the regulations can be tightened up from time to time in the light of technical progress that may be made in pollution-abatement techniques, for the whole point of the charges method is that it provides a continuous and permanent incentive to find improvements in such techniques. Technical progress in reducing pollution will therefore be much faster with a pollution charge than with direct regulation. Consequently, the amount of pollution will tend to be lower.

For various reasons, therefore, pollution charges (or some other price-mechanism system)[2] will enable society to reduce pollution more cheaply than direct regulation. Firms will be better placed to find the cheapest method of reducing pollution, pollution reduction will be concentrated on the firms that can reduce it most cheaply, and technical progress in pollution abatement will be continually stimulated. And if pollution charges enable society to reduce pollution more cheaply, it

[1] There are numerous examples of cost-saving technological progress in pollution abatement, such as those given by Paul Gerhardt (chief of the Economic and Science Studies Section, National Center for Air Pollution Control, US Public Health Service), in 'Incentives to Air Pollution', *Law and Contemporary Problems*, Vol. 33, No. 2, 1968.

[2] E.g., subsidies related to the amount by which pollution is reduced; compensation of the victims by the polluters, etc. (above, pp. 49-50).

follows that either a given amount of abatement can be achieved at less sacrifice of other goods and services or that, for the same cost, pollution can be further reduced.

(iii) *Consistent, automatic application*

The lower cost of pollution abatement under the pollution-charge method is not, however, the only advantage of this method over that of direct controls.[1] A third advantage is that direct controls tend to be uneven in their application according to how popular the anti-pollution fashion happens to be. At a time when the environmental issue is front-page news and any case of an excess of some pollutant entering the environment hits the headlines, some of the authorities responsible for pollution control may exercise more vigilance. But a few years later, when it is education, or the health service, or housing, or public transport, or crime, or drugs, or something else that happens to be the prime concern, environmental protection may not be enforced with quite the same enthusiasm.

This is in no way a criticism of the local authorities or other bodies to whom responsibility for environmental protection has been delegated. But the resources and funds put at their disposal are obviously a function of how far their political masters, whether they be locally or centrally elected, think that pollution is a 'hot' issue. Furthermore, the importance attached to protecting the environment, which may often appear to be at the expense of other local interests, such as employment in some polluting industry, may lead to local or regional variations in the enthusiasm with which anti-pollution policies are pursued.

(iv) *Inefficiency of direct controls*

Finally, the enforcement of direct controls is often a difficult and time-consuming process, requiring, for example, the accumulation of sufficient evidence to satisfy a court of law that a polluter has exceeded the limits laid down; even then the fine imposed is often derisory. Indeed, direct regulation is also a form of pollution tax, in the sense that a small fine may be imposed if breaches of the regulations are identified and

---

[1] The following few paragraphs owe much to the points made by Professor W. J. Baumol in his 1972 Wicksell Lectures, *Environmental Protection, International Spillovers and Trade*, Almqvist and Wicksell, Stockholm, 1971.

proved to the satisfaction of the courts. But the incidence of this form of tax is often uncertain, subject to delays, and usually too small anyway.[1]

By contrast, no such vagaries apply to the operation of a charge or tax. Once a tax is instituted a proper machinery has to be set up (and is invariably set up) to ensure that the required data—however rough and ready they may often have to be—are provided as frequently as required, taking account of feasibility and so on. The collection of the charge or tax is then a routine matter unaffected by changes in the winds of fashion or local pressures. Tax collectors collect their taxes year in, year out, and in the same way from one part of the country to another. In fact, a system requiring regular returns of liability to a pollution charge would be one way of increasing our information about pollution, hence making it easier to determine optimum pollution levels. Furthermore, it would probably stimulate technical improvements in monitoring techniques, such as those that have recently been made in response, no doubt, to the emerging interest in pollution control.[2]

V. POLLUTION CHARGES VERSUS DIRECT CONTROLS[3]

In view of these four apparent major advantages of the pollution-charge system over direct regulation, one may ask why the opposition to it is so widespread. Yet one of the most hotly disputed issues of pollution policy today concerns not the choice between alternative forms of 'tinkering with the price mechanism' but between *any* such method, on the one hand, and some form of direct regulation of pollution, on the other hand. Various reasons have been advanced for preferring

[1] In Britain over the whole period 1967 to 1971 inclusive, there were eight convictions for air pollution as a result of prosecutions by the Alkali Inspectorate, and the average fine imposed was £3. (Reply by Mr Eldon Griffiths to Parliamentary Question, 3 May 1972.)

[2] Cf. reports on major technical advances in monitoring equipment in the *Financial Times*, 12 October 1971, and the *Guardian*, 25 June 1971, both of which related to the development of much more automatic instruments for measuring various parameters of water quality.

[3] This whole section follows very closely the discussion of this issue in the Minority Report by the present writer and Lord Zuckerman contained in the Third Report of the Royal Commission on Environmental Pollution (Cmnd. 5054, HMSO, London, 1972).

direct controls to some price-mechanism instrument to check pollution, and the more common ones will be dealt with in detail below (pp. 59-67).

Ideological preference?

It should be conceded at the outset, however, that a preference between a price-mechanism scheme and direct regulation may reflect a preference between the price mechanism and direct regulation as a general system for achieving the desired allocation of resources in the economy, subject to some fairly clearly defined exceptions. In other words, a preference for direct regulation of pollution is rather like a preference for direct regulation over free collective bargaining as a means of obtaining the desired allocation of labour, or over the free market for the desired production of shirts.

A pollution charge does not constitute some completely novel or revolutionary scheme, it is simply a time-honoured price mechanism. Indeed, in the past it has been the 'revolutionary' régimes that have tended to replace the price mechanism by direct regulation, and many of the people now opposed to some sort of pollution charge would probably be surprised to discover what ideological camp they are really in.

Approach to pollution charges

Nor do pollution charges constitute some novel, untried method dreamt up by ivory-tower economic theoreticians. In Britain, for example, some form of financial disincentive to pollute already exists, up to a point, in the charges for the treatment of 'trade' (that is, non-domestic) effluent discharged to municipal sewers. The Public Health (Drainage of Trade Premises) Act, 1937, and the Public Health Act, 1961, provide drainage authorities with adequate authority to control the discharges into their sewerage systems and to

'charge for the reception of the trade effluent into the sewer regard being had to the nature and composition and to the volume and rate of the discharge of the trade effluent so discharged . . .' (Public Health Act, 1961).

A large number of local authorities charge for trade effluent according to formulae which take account of some indicators of pollution (notably the BOD[1] and the amount of suspended

---

[1] 'BOD' (biochemical oxygen demand) is one measure of the oxygen, in the water, that is used up by the matter contained in it. The higher the BOD, therefore, the less oxygen available to support fish-life.

solids in the effluent). Furthermore, various bodies and individuals concerned with the practical application of charges have confirmed the feasibility and effectiveness of a charges scheme. For example, the Institute of Water Pollution Control stated many years ago:

> 'One of the most effective methods of reducing the load caused by trade effluents is to make a charge for their treatment which is based on a sliding scale in accordance with their volume and strength. In this way an incentive can be given to the trader to reduce his discharge of waste from his factory, by re-using water, by making minor modifications in manufacturing processes, by recovering by-products or by some other means. Some remarkable results have been achieved in this way with profit to the trader, and with great advantage to sewage works operation, and with considerable resulting contribution to the national economy.'[1]

American experience

There has also been a move towards the use of pollution charges in the USA, beginning with the imposition of a charge on sulphur oxide emissions announced in the Presidential Message to Congress on 8 February, 1972. This charge, to operate as from 1976,

> 'is an application of the principle that the costs of pollution should be included in the price of the product. Combined with our existing regulatory authority, it would constitute a strong economic incentive to achieve the sulphur oxides standards necessary to protect health, and then further to reduce emissions to levels which protect welfare and aesthetics'.

Other moves in the direction of taxing pollution or similar externalities are under consideration, such as the US Administration's proposal to tax lead additives in gasoline,[2] and the

---

[1] *Memorandum on National Policy on the Discharge of Trade Effluent into Public Sewers*, by the (then) Institute of Sewage Purification, 1952. The same advantages for the charging system have been claimed in a paper by Mr Goodman, Chemical Inspector, Directorate of Water Engineering, Department of the Environment, submitted to the Economic Commission for Europe, January 1972, p. 2. See also Simpson and Truesdale, *Methods of Charging for the Treatment and Disposal of Industrial Effluent in Municipal Sewerage Systems*, paper presented to the Institute of Water Pollution Control Symposium, London, November 1971.

[2] *Report of the President's Council of Economic Advisers* (USA), 1971, p. 119. However, not much progress has since been made in implementing this tax.

Japanese Government's consideration of a congestion tax on firms in congested areas.[1] And recently, the notion of taxing pollution received support even from the International Chamber of Commerce at its meeting in Goteborg in 1972.[2]

In the USA, unlike Britain, the environmentalists seem to have woken up, for one reason or another, to the idea that a pollution tax or charge is more likely to meet their aims than direct regulation. It has been reported that

'. . . over the past year organised environmental groups have gradually come to see a pollution tax as a strong incentive to institute abatement measures. They also cite the fact that most industrialists have declared their opposition to the measure as proof that it would force them to take the painful steps to reduce their sulphur emissions. A measure of the environmentalists' interest in the concept is the fact that in August last year a number of groups joined to form a "Coalition to Tax Pollution" and set up an office on Capitol Hill'.[3]

Arguments against pollution charges

Numerous detailed arguments against pollution charges have been put forward from time to time, and it would take too much space to cover them all here. But the following seven arguments seem to be a fair sample.

(i) Charges more expensive than controls

The first, and in a sense the most serious, objection is that polluters will, in practice, find it *more* expensive to reduce pollution by a given amount, since, in addition to the *real* costs of abatement they will incur in order to reduce pollution (i.e. the additional labour, capital, and raw materials they

---

[1] *Financial Times,* 11 August, 1972.

[2] 'Taxing Polluters', in *Marine Pollution Bulletin,* Vol. 3, No. 7, 1972.

[3] 'Pollution Tax', *Nature,* 18 February, 1972. A similar report was contained in the *Guardian,* 14 July, 1971. The environmentalists are still far from understanding the problem fully, however, judging by their objection to the differentiation in the proposed US sulphur-oxide tax on the grounds that this will tend to lead to a transfer of pollution from highly congested areas where it does a lot of harm to other areas where it will do little harm, and hence where the tax will be low or negligible. But such a transfer is, of course, desirable, in the light of the principles set out above to the effect that pollution *per se* is not important; what matters is the damage it does, and such a transfer clearly reduces the total damage. Would the environmentalists complain, for example, if it could all be shot out into space at no cost, so that nobody was harmed by it at all?

have to substitute for the environment), they will also have to pay the pollution tax on their residual pollution. By contrast, it would be argued, if they are obliged to reduce pollution by direct regulation they incur only the real costs of abatement. If the latter are higher, because of the failure of direct controls to allocate abatement among polluters in line with their relative costs of abatement, this would be offset by the extra burden on polluters embodied in the pollution tax which they have to pay in addition to their abatement costs.

The fallacy in this argument is that we want to know which method will be cheapest for the economy as a whole, other things being equal, notably the general level of demand and employment. For we are interested in the effects of resource re-allocation, not in the effect of reducing the amount of resources used as well. Of course, if the authorities deflate the economy by raising some taxes that are not neutralised by compensating tax reductions elsewhere (or compensating increases in public expenditures), total output and employment will be reduced, so that the economy as a whole could be worse off than if some other method had been used. But the costs of the pollution abatement then consist of both the real resources used to abate pollution in the firms concerned *and* the fall in output and employment as a result of the unintended and unnecessary deflationary effect of the higher tax receipts of the public sector as a whole. But this would no longer be a comparison of the relative costs of the alternative methods of pollution abatement, other things being equal, notably the general level of demand and employment. Clearly, to make the correct comparison, it is necessary to assume that the authorities neutralise the demand impact of their pollution-tax receipts by either tax cuts elsewhere or increases in public expenditure elsewhere. In that event, the only cost of the pollution abatement is the total cost of real resources used by firms to reduce pollution; we must exclude the pollution charges that are also paid since they, or rather their effects, must be assumed to be offset by neutralising changes in the government budget.[1]

Indeed, if the compensating action by the authorities takes the form of tax reductions, there may be a yet further gain in efficiency in the economy. For most taxes lead to a misallocation of resources. A pollution charge is an exception; it corrects a

[1] From the point of view of the economy, taxes are transfer payments, not resource costs.

[60]

resource misallocation caused by the externality aspect of pollution. If the revenue from pollution charges enables the state to reduce other taxes it will tend to reduce the resource misallocation that is caused by these other taxes and hence make a further contribution to the optimum allocation of resources. This constitutes yet a further way in which pollution charges have a cost advantage over direct regulation.

(ii) *'Unrealistic assumption' of profit maximisation*

A second objection to pollution charges is that the theory, as advanced above, is based on the unrealistic assumption that firms are ruthless profit-maximisers, making careful calculations of the optimal degree to which they should reduce pollution in response to the charge.

The case for pollution charges no more rests on this unrealistic assumption than does the case for, say, the use of the price mechanism to allocate labour or capital investment between firms. Of course firms are not at all like this, and most firms do not make the theoretically ideal calculations of their investment needs, for example, but even those who are opposed to pollution charges would not argue, on this account, that firms' investment projects should be determined by direct regulation. Hence, the above argument no more depends on a very simple view of the way that firms operate than would arguments in favour of using the price mechanism rather than direct regulation to allocate labour, capital and raw materials between different firms according to some quantitative plan.

(iii) *Ineffective—higher prices*

A third and related objection to pollution charges is that they can have no effect since the polluters will merely pass them on in higher prices.

Producers normally try to cover *all* their costs in their prices —otherwise they would soon go out of business, yet it does not follow that they are indifferent to how much labour or capital they use.[1] Firms will still try to employ each factor of production

---

[1] Similarly it is absurd to argue, as did the representative of the Confederation of British Industry (Mr Biggs) at a conference on the environment in May 1972, that it would be easy for companies in a monopoly position to pay the tax rather than cut pollution. It is no more easy for them to do this than to pay more wages instead of economising on their labour force. The CBI does not argue that monopolistic firms should be directed as to how much labour they use, rather than just pay a wage, on the ground that they will simply pass on the wages in higher prices.

up to the point where further use would not add to their revenues more than they add to their costs. In general, it is the more profitable firms that carry out this process more efficiently. To assume that by paying a 'licence to pollute' firms have no incentive to economise on pollution is like assuming that firms do not economise in their use of other factors of production.

If this were true the whole allocation of resources in the economy would be completely haphazard. It should not be forgotten that a pollution charge would not be like a radio licence which, once paid, entitles the licence-holder to an unlimited amount of listening. A pollution charge would be related to the amount of pollution; the more one pollutes the more one pays. In any event, as the 1971 Annual Report of the President's Council of Economic Advisers pointed out,

'. . . every system of rules for use of the environment, other than outright and total prohibition of certain uses, involves granting someone the right or "licence" for some polluting'.

(iv) *Unworkable—data deficient*

It is often maintained that charges are unworkable because we do not have the data required to decide on the appropriate charges in all cases. We do not, for example, have the data required to make full allowance for variations in the conditions (such as state of river flow, air temperature, tidal conditions) that determine the amount of damage done by effluent at any point in an estuary.

This is true, but the same lack of data means that one does not know the correct amount of pollution abatement to be imposed by direct regulation either. In principle, the data required to identify the correct objective in pollution abatement are independent of the methods that might then be used to achieve it. For these data are the same in both cases: they consist of the costs of abatement and the damage done to society by the pollution, and are therefore the same whatever policy instruments are to be used for abatement purposes. A person might not, for example, know whether he will have more sunshine on his holiday if he takes it in Scotland or in Brighton; and he may also have a choice of going to either place by rail or by car. It would be irrational for him to say 'Since I am not sure which place is best, I shall go by car'. It is not by going by car that he will increase his chances of guessing correctly where he will find the most sunshine. The two issues—the correct objective and the means of achieving it—are quite separate, as they are with pollution abatement.

Moreover, for the administrative reasons set out above, it is likely that these data would be built up more rapidly if some authority were responsible for regularly levying charges based on the amount of pollution than if pollution control were left to direct regulation.

(v) *Impracticable—measurement difficulties*

It is claimed that charges are impracticable because data are not available to permit an accurate calculation of the amount of pollution that should be taxed. Monitoring difficulties may, for example, preclude the observation of the pollution which is to be charged.

This is true, but, again, precisely the same problems apply to the surveillance and implementation of direct controls. The imposition of direct control implies that whatever is controlled can be measured—otherwise it is pointless to institute the control, since it would be impossible to check whether it is respected. And whatever can be measured can also be taxed. If a firm is instructed not to put more than 1 ounce of some heavy metal per day in the river it is unlikely that the check, in so far as there is one, on the amount of the metal contained in its effluent indicates only whether the amount discharged is above or below 1 ounce. A more informative, if not precise, figure would usually be obtained, such as that the amount was 2·8 ounces or 0·7 ounces. The extra information would be virtually free in most cases, and would be adequate for a charging scheme as long as it was recognised from the outset that the charging scheme would be no more precise than the direct control.

Even where it is not practicable to measure how much pollutant is in the effluent and it is necessary, if any control at all is to be exercised, to lay down consent conditions in terms of, say, the raw material or the productive process, this will usually still be related to some quantifiable flow or characteristic of the raw material or productive process which could then be used as the basis for a charge. Suppose it were thought desirable to reduce the amount of a heavy metal flowing into an estuary, but it was impossible to measure the metal at the low concentrations that might be relevant. Direct control, if any, might then take the form of a restriction, in some productive processes, on the use of a raw material believed to be responsible for the pollution. But if the amount of the raw material used can be measured to ensure that the direct control is respected, the measurement can equally be used as a basis for the pollution charge.

Similarly, if a minimum height of a chimney has been stipulated it must be possible to obtain a measure of the chimney height in order to check that the regulation is being obeyed. It would cost no more to use the measure for purposes of a tax that varied (downwards) according to the height of the chimney. In some cases, of course, verification difficulties may mean that the calculation of the charge will be inaccurate, but the check on the observance of direct controls would then be equally unreliable. In other cases, the costs of operating a charging system might be excessive in relation to the damage done by the pollution, and it might then be thought not worth while. But it would probably be equally undesirable to attempt to monitor how far the direct regulations are respected.

The whole argument may perhaps be illustrated by considering an apparently absurd and trivial case: the offensive smell of fried onions from a restaurant. Suppose it were decided that this smell must be reduced, but that it was impossible to measure smell,[1] so that clearly it would also be impossible to tax it. But it would then be equally impossible to limit the amount of smell by direct regulation. For there would be no point in the Inspector saying that the smell from the restaurant was too strong that day, since the owner could hotly deny it and there would be no objective means of settling the dispute. One could ask a lot of people to come and sniff, but they would not know exactly what standard of smell the Inspector had in mind. In such a situation direct control might take the form of imposing a quantitative limit on how many fried onions the restaurant was allowed to use per day, assuming that the Inspector had some means of controlling this—i.e. of measuring the quantity of onions used every day. But if this assumption

[1] But even the measurement of smell should not be thought of as beyond the realms of possibility. As a result of much experimentation in Sweden it now appears that 'Through an instrumental technique . . . it has proved possible to measure objectively the concentration of evil-smelling sulphur compounds with a precision and sensitivity comparable with those of the nose', according to the *Bulletin of the Swedish Water and Air Pollution Research Laboratory*, Vol. 1, No. 1, 1972. Similar progress, but with a different technical approach, has been reported in Britain, in *Pollution Monitor*, August/September 1972.

is made it is obviously equally feasible to tax the use of fried onions. Again, one returns to the basic proposition: *whatever can be controlled must be measurable; if it is not measurable it is an illusion to believe that it is being controlled; and if it is measurable it can be taxed or priced (charged for).*

(vi) *Inadequate control via charges*

It is often believed that the major advantage of direct regulation is that the regulating authority knows exactly whether or not the abatement target will be achieved, whereas with a charge system they will not know in advance how far firms will respond to the charge and hence how far pollution will be reduced to the optimum amount.

This is very much like arguing that the big advantage of direct regulation of clothing output in centrally planned economies is that they can be sure that the target for clothing output will be produced, whereas if they had left it to the market mechanism clothing output might have fallen below or above the target. This is quite true, but the accuracy with which one hits any target is not, in itself, a desirable objective of policy, irrespective of the extent to which it is the *appropriate* target. The advantage of the price mechanism is precisely that if the output of clothing is too high its price will fall, thereby discouraging its production (and encouraging its consumption) until the correct amount is produced. But with a production quota, and, in addition, no market (as would be the case with pollution), producers would continue to produce the target level of pollution and nobody would know whether or not it was the *correct* target. By comparison, with the charge method, if the charge failed to produce the level of pollution at which the marginal social damage were equal to the charge, this would itself constitute evidence for supposing that the initial estimate of optimum pollution could not have been correct, and that an adjustment in the charge would bring the optimum nearer.

(vii) *Lack of precision*

A seventh argument often advanced against charges is that, while polluters will accept rough-and-ready measurements for purposes of direct control, only very precise measurements will be publicly acceptable if they are to be the basis of a tax.

This is manifest nonsense, for innumerable taxes, fees, dues, prices and charges of one sort or another are levied in a rough-

and-ready manner without giving rise to any general refusal by the public to pay them. Local rates are not only calculated on very rough-and-ready formulae for rating valuations; they are not even revised very frequently to allow for changing circumstances. Charges for bus fares, telephone calls, parking meters, and innumerable other services where it would be relatively easier to adjust the charge according to some very finely graduated scale are, in practice, arranged according to a scale with relatively large steps. With most pollutants it would be even more desirable to vary the charge according to relatively large steps in the pollution load.

Quite apart from the difficulties of precise monitoring (which apply to any system of control), the damage done by any amount of pollution varies considerably according to many other physical parameters, such as the composition of the effluent, the air conditions, the river flow, the time of day or night, and so on. It would be foolish to try to be any more perfectionist about a charging scheme than about direct control schemes.[1]

Related to this is the argument that a pollution charge requires continual monitoring. But a charge scheme no more requires continual monitoring than does a direct control scheme. If the direct control takes the form of a weekly check on the flow of a firm's effluent to ensure it is within the consent limit, the same readings can be used as a basis for the charge. It is absurd to argue that, with the charge system, the firm can always seek to cheat by disposing of its pollution at some other time of day or week, when the flow is not being measured for the charge purpose. For it could do exactly the same to avoid being caught when the flow is being measured for the purposes of checking whether it is respecting the direct regulations. Thus, it is not true that the use of the charges scheme depends on technical improvements in monitoring arrangements. Such improvements will make it easier to control pollution by any means, and are irrelevant to the choice between the direct control and other methods.

There are, of course, many practical problems in introducing a charge scheme, such as the role of the various authorities in

[1] It is not widely realised that the current practice of concentrating on the content of SO_2 and smoke, as far as air pollution is concerned, is largely a matter of convenience in that these two particular pollutants are regarded as being very good indications of air pollution in general, as well as being pollutants in their own right.

determining the target amounts of pollution in each case, in setting the charges, and so on.[1] It may often be too costly, in relation to the benefits, to try to implement a scheme. But it is more important here, where we are concerned with basic principles rather than their detailed individual implementation, to move on to other major items concerning the role of the public authorities in pollution control. These include the relative contribution of collective purification facilities, such as sewage works, as against pollution charges, as a means of reducing pollution, and the relationship between the revenues from pollution charges and the number of public purification facilities that should be provided.

VI. COLLECTIVE PURIFICATION FACILITIES AND 'PUBLIC GOODS'

Economies of scale

One important alternative to pollution charges or direct controls as a means of reducing one of the most common forms of pollution, namely water pollution, is to allow firms (and households) to pour their effluent into a collective drainage system and then purify the water to the desired degree in a collective sewage works. It might be much cheaper to do this when there are economies in treating effluent on a large scale instead of obliging each firm or household to install a small-scale purification process. (This applies chiefly to water pollution; there are few such possibilities in most other forms of pollution.) But it is wrong to regard collective facilities and pollution charges as mutually exclusive alternatives. In an optimal system for controlling pollution both have their part to play, together with many other ways of reducing pollution, according to their respective cost conditions.

[1] Some of these ideas are discussed in the Third Report of the Royal Commission on Environmental Pollution, *Pollution in some British Estuaries and Coastal Waters*, Cmnd. 5054, HMSO, London, 1972, pp. 82-5. I do not cover, in this *Paper*, some of the more absurd anti-charge arguments, such as those put forward in an editorial in *Pollution Monitor*, October-November 1972, to the effect that the charge scheme is undesirable because it would require such a wide variety of charges. This is like arguing that the allocation of motor-cars by a price mechanism is undesirable because there is such a wide variety of prices. Anyway, the direct regulation system also involves extensive variation from firm to firm in the composition and quantity of the pollution they are allowed to produce.

The supply of clean water from collective sewage, or other purification facilities, together with the supply of clean water determined by the amount of pollution produced in individual sources, constitutes the total supply to the community, and its price should be the same for all users and to all suppliers. Individual firms will have to pay for using the collective facilities to purify their effluent, and this charge should correspond to the tax they would bear on their pollution. In this way they will tend to find the optimal allocation between reducing their pollution themselves, paying for the collective facilities to do so, and paying a tax on their remaining pollution.

The case for collective facilities to deal with pollution does not rest solely on possible economies of scale. Whether collective facilities are under public ownership or not is nothing to do with the existence of economies of scale,[1] and the collective facilities could be privately operated, as are, for example, many water-works. Thus the provision of collective facilities on the ground of economies of scale is a separate issue from the provision of publicly-operated collective facilities on the ground that they provide what are known as 'public goods'.

'Public goods' criteria

There are various criteria by which activities are classified as 'public goods'. One of the most important is that provision of the good or service to one person, or many people, automatically makes it available at no extra cost to other people. The classic textbook example is the lighthouse, where, as long as the light is made available to one ship it can be seen, at no additional social cost, by other passing ships, at least up to the point where the area concerned becomes congested. National defence is another example: once the army or navy is established to defend some of the people, the others are equally defended. The two examples are not identical, since some of the defended people might not want to be defended at all, but they will be just the same, whereas only sailors that *want* to look at the light need do so. But the common feature is that, once the service is provided for some users, society need incur no additional opportunity cost in order that others may enjoy it. Hence, it would be sub-optimal to charge any

---

[1] Apart from the usual problems of optimal pricing policy in industries where average costs of production decline as output rises.

consumer for the use of the service in question, since such a charge would merely reduce his consumption and welfare without adding to the possible consumption of anybody else.

At the same time it is usually impossible to prevent anybody from using a public good—for example, to prevent any particular ship from taking navigational bearings from the lighthouse. This means that it is usually impossible to charge anybody for using it. These two features of public goods mean that the free-market mechanism is hardly likely to produce the socially required amounts of them. The public sector consequently has to step in and fill the gap.

Pollution a public 'bad'

Pollution is a form of public 'bad'. An individual who breathes polluted air, or smells a polluted river, does not usually reduce the amount of polluted air or smell available for other people. Conversely, clearing up pollution is a form of public good: reducing health hazards from poor sewage for some people in any given locality will, at the same time, reduce the health hazards for other people in the same area or for visitors. Hence in the same way that the public-good character of, say, lighthouses, or defence, implies that the price mechanism cannot ensure the socially optimum output of the service in question, it might appear that the public-good character of, say, a sewage works or other pollution-prevention device necessarily implies that it must be supplied by the public sector. But this is not so: the public-good character of pollution prevention and purification is not completely analogous with the classic public good.

The difference is that with, say, the lighthouse, the only way that the dark can be mitigated is by building the lighthouse, whereas with *currently* produced pollution there is always the alternative of taxing or charging the polluter. Currently produced pollution is something that can be *reduced at source*, whereas this is not so with, say, darkness or the hostility of an enemy country (real or imagined). Only if darkness could be reduced by taxing God might this be preferable to building a lighthouse.

Since this reasoning does not apply to *previously* produced pollution, the case for a 'public-good' approach to, say, the restoration of derelict land, or beaches that have been polluted by some previous oil-spillage at sea, is much stronger. But with currently produced pollution the case for public facilities is by no means conclusive: the optimum amount of pollution abatement can often be achieved by a pollution charge, supplemented by collective facilities (possibly privately operated) when there are economies of scale.

Even with pollution charges, the market mechanism may still fail to ensure that the socially most economic system of reducing pollution is introduced. There are too many imperfections in the market, as well as the expenses involved in obtaining the requisite information or in conducting the appropriate transactions. Hence, the appropriate collective facilities that would be required for an optimal solution may simply not be constructed by anybody, so that the public authorities will often be obliged to intervene. But where they do so, they should still charge for their service as if they had been some large-scale commercial firm supplying 'clean water' (or other medium) to those who found it cheaper to obtain it this way than to produce their own.

Who pays and what happens to the money?

Instead of the polluters, would it not be preferable to make the beneficiaries pay? After all, it may be argued, in so far as some people benefit from a cut in the pollution of, say, water, should they not be charged for the clean water in order to discourage them from wasting it? Why give them a free gift?

It is perfectly true that, in so far as public authorities can charge for any clean product they supply as a result of their pollution-abatement or purification facilities, and in so far as these do not have a pure 'public-good' character (notably that the more any person uses of it the less is available for others), a charge should be made, as, for example, with the supply of piped drinking water, or access to special recreation facilities that would otherwise become congested (thereby involving an opportunity cost). In the same way that a firm will waste the pollution 'input' if it is free, consumers will waste clean water if it is free.

But this does not mean that the polluter should not be charged; one does not exclude the other. The price of clean water should be the same for all *users*, whether the user is the consumer who washes in it, the factory that uses it for industrial purposes, or the factory that pollutes it. Charging those who benefit from the purification of the water by no means implies that polluters should not also pay for their pollution. Both should pay, for both are, in effect, users of the clean water, and only if they face the same charge will the optimum total supply of the clean water be obtained and allocated between them in the optimum manner. Apart from the pure public good, all users of a scarce clean medium should pay for it, whether they use it as final consumers or destroy it through their pollution.

Another minor point concerning charges for public facilities is the question of how far the revenue from the charges should determine the amount to be spent on environmental improvement (as with the French Agence des Bassins system). There should be no connection between the amount received from pollution charges (or from the provision of clean water) and the amount spent on purification or other environmental protection. How far the public sector reduces pollution, or cleans it up, should be determined by the principles set out above—i.e. the relative cost and feasibility of collective facilities, or the 'public-good' character of pollution treatment (notably where it is a matter of cleaning up past pollution). It has nothing to do with the revenue that would be derived from either pollution charges or charges to users.

These revenues are relevant only in connection with their impact on the general level of demand, as discussed above (pp. 59-61). That is to say, the revenues from pollution charges must be offset by reductions in other government tax revenues (or increases in government expenditures) in so far as the government wishes to maintain the same pressure of demand in the economy. But which taxes should be reduced or which alternative government expenditures should be increased should bear no relation to the origin of the additional pollution charges.

VII. CONFLICTS BETWEEN ENVIRONMENTAL POLICY AND EMPLOYMENT OBJECTIVES

So far we have been concerned with the principles of pollution abatement only from the point of view of the optimum national allocation of resources. We have examined the way in which these principles would be served by policy instruments that, in one form or another, make the polluter bear the full social costs of the pollution for which he is responsible.

Economic policy is designed to serve more than one economic objective—though perhaps not quite as many as is often believed. One of the most important objectives during the last few decades or so has been to maintain full employment, both nationally and regionally, i.e. to ensure that resources are fully used, rather than to worry about *how* they are used. Governments are also concerned with income distribution, i.e. with the way the output produced by all our resources is allocated between people. And this is often closely bound up with the impact of policies on local employment, or on specific industries or sectors of the economy.

Regional policy

Political concern with regional differences in prosperity may well be as much a reflection of the political constraints on all governmental policies as a concern with general economic welfare. For it is not obvious why, other things being equal (including duration of unemployment, and so on), total economic welfare is reduced if a given total number of unemployed tend to be concentrated in one region or one industry rather than spread over the country. But in the former case the political pressures on governments to act are likely to be far stronger. Hence, there may often be instances where a reduction in pollution in the interests of better resource allocation appears to conflict with other objectives, in that it would have a particularly damaging effect on employment in a locality or an industry, or that it would hit old and/or small firms in particular.

Protection-of-industry conflicts

One very obvious example of a frequent conflict between the optimum allocation of resources nationally and a possible loss of welfare (through loss of jobs) for those engaged in a particular industry or region arises where decisions have to be taken on the protection of industries against foreign competition. The pure theory of international trade might demonstrate that a reduction in barriers to trade is desirable in the interests of world welfare, or even of the welfare of individual countries, but the theory nowhere suggests that it is always necessarily desirable in the interests of the welfare of all individual groups. A reduction in tariffs on some imports might raise national economic welfare if the gains to the consumers outweigh the losses to the producers (which will usually be true).

What is to be done about the losers? It is little consolation to them to be told that the economy as a whole is better off and that it would have been possible, *in theory*, for them to be compensated for their losses by the consumers, leaving the latter better off than they were to begin with. The same principle applies if individual industries face higher costs as a result of some need to reduce pollution. Governments will obviously be under pressure not to introduce anti-pollution measures that could create difficulties of this sort. Pollution abatement would not impose a heavy burden on the country because of a loss of competitiveness in international trade.[1] But this does not dispose of the problems that could arise for the individual industry, firm or region. And here governments will face a conflict of objectives.

Three solutions

Broadly speaking, there seem to be three types of solution to adopt in the face of such a conflict. *First*, the measures that would be appropriate on resource-allocation grounds can be modified or relaxed; *secondly*, they can be maintained, but with a time-lag to allow for a transitional period of adjustment; *thirdly*, they can be implemented without qualification, but accompanied by additional measures designed to minimise the conflict with other objectives.

First, suppose the appropriate policy for pollution abatement meant that considerable extra costs were imposed on some firms or an industry in a particular area, with the result that their competitive position was badly threatened (nationally or internationally) and considerable local unemployment ensued. Many authorities would be under pressure to abstain from the appropriate anti-pollution measures and replace them by others that would be less efficient, from the point of view of resource allocation, such as a subsidy to the industry to install techniques of production that involved less pollution. This would be the first type of response to the conflict of objectives. It is open to the usual objection that, in so far as resources are misallocated, total national output is less than potential output. If, instead, output were maximised (by

---

[1] The case is argued in Chapter 7 of *In Defence of Economic Growth*.

optimum resource allocation) it would, at least in principle, be possible for the losers from the anti-pollution policy to be compensated, or more than compensated, while the rest of the community would still be better off. Whether the losers would, *in practice*, be compensated is another matter, and would depend on social and political circumstances. While economists may have no expert knowledge of these circumstances, it is important that they draw attention to this aspect of the problem, rather than give the impression that there is absolutely no reason known to economic science why the resource allocation objectives should ever be sacrificed in the interests of income distribution.

The *second* type of response to the conflict of objectives is to permit a transitional period during which firms have time to take appropriate measures to adjust themselves to the introduction of pollution-control policies. This is the common practice whenever tariffs on internationally traded goods are reduced. The internationally agreed rounds of tariff reductions and also the arrangements for the establishment of customs unions of one kind or another (such as the EEC) invariably allow transitional periods. The *rationale* is usually that it is inequitable to remove protection from domestic firms suddenly in the interests of resource allocation, since the growth of the industries, the investment of capital and the acquisition of skills and other ties by the labour force therein have been developed when tariffs did exist. Hence, workers must be given time to find other outlets, other job opportunities, or other ways of adapting themselves to the changed market conditions, which have been brought about through deliberate governmental policy rather than through the normal uncertainties of economic life.[1]

A *third* reaction is to implement the policy to restrict pollution to the (assumed) optimum and to deal with the other problems that may then arise by entirely different measures. It might be thought that the best procedure would be to implement the pollution-abatement policy and accompany it by measures to improve labour mobility, or re-training, or to inject alternative sources of employment into the region. In some cases the accompanying measures might involve even

---

[1] While profit-receivers might be expected to bear the risks of uncertainty on the ground that this is what profits are the reward for, it is difficult to justify the notion that labour should bear any of these risks.

more resource misallocation of a new form than did the initial excessive pollution. Measures to stimulate artificially the entry into a region of new industries that do not, otherwise, find it economic to go there will often result in resource misallocation, unless they can be justified in terms of dynamic effects in the long run, or additional economies of scale, and so forth. On the other hand, some measures to minimise the local-employment effects of anti-pollution policies might only involve improvements in information, at little cost, which would thereby increase the efficiency of national resource allocation.

In principle, this third means of reconciling conflicts in policy objectives is likely to be the most efficient economically.[1] If the objective of reducing pollution conflicts with the employment objective it is generally preferable to persist with the pollution abatement with the aid of the most appropriate instrument (namely, a pollution charge), but to accompany it by an additional instrument, such as a fiscal or monetary or institutional instrument, designed to bear on the level of employment.

Local unemployment and pollution costs

Apart from recognising that, in principle, the third response to the conflict of objectives is the most preferable, it must be acknowledged that, in practice, the appropriate instrument may not be at hand, particularly when it is local employment that is affected. Where pollution abatement means a loss of jobs and the only immediate alternative is unemployment, the true cost to society of labour employed in the industries concerned is less than the wage—i.e. it is less than the nominal market price of the labour. This is because the use of labour in those industries does not imply that it has been taken away from some alternative activity and that society is thereby deprived of the output of this alternative activity. As explained earlier (pp. 39-40), what matters for resource allocation are 'opportunity costs', i.e. what output is sacrificed in one part of the economy as a result of using resources elsewhere. If the labour would *not* have been employed in some other way the social 'opportunity cost'—which is the true social cost—of using it in the polluting industry is nil.

---

[1] This corresponds to the well-known principle of economic theory that it is impossible to achieve several targets successfully without an equal number of policy instruments (that independently affect the different objectives).

In such cases, the resource misallocation from excessive pollution, in the short run, is reduced and might be zero, or negative. For, although the nominal market costs of the goods concerned fail to allow for the external costs of the pollution generated in their production, they exaggerate the true social cost of the labour employed. In such a situation, the second response to the conflict of objectives might be appropriate, namely the use of a transitional period, accompanied, as far as possible, by measures to minimise the transitional difficulties, provided they did not tend to perpetuate the resource misallocation.

On the whole, this procedure would appear to be preferable to the first type of response, such as subsidies to firms for pollution-abatement equipment. In general, such subsidies are not likely to be very effective, except where accompanied by other measures to enforce or stimulate pollution abatement, and will anyway not lead to the most economic means of pollution abatement, as would be the case with the pollution charge. They are also likely to be diverted partly to subsidise investment in general (though this side-effect may not always be entirely undesirable). While economists are reluctant to stick their necks out over questions such as whether any particular policy is necessarily desirable from a welfare point of view, they do have a duty to draw attention to some of the fallacies behind many of the anti-pollution-abatement arguments.

Pollution abatement and unemployment not alternatives

That pollution abatement might mean a loss of jobs, for example, does not necessarily mean that it should not be adopted; for, as pointed out (pp. 73-75), it is at least necessary to be sure first that no other means can be adopted to remedy the employment problem. If no such policy can be devised, so that a chronic, quasi-permanent, increase in local unemployment would result, that is another matter. But the relevant facts must be established in the first place. The unemployment argument has been used throughout the ages to oppose all sorts of measures to reduce international tariff barriers, to introduce safety regulations or other improvements in working conditions in factories, to abolish child labour, and so on. But in the end these measures have been adopted; some short-term local effects on employment may have been felt (and sometimes the effects were acute and long-lasting), but this has not led to increasing unemployment in general; and, in the long run, standards of living have risen.

Thus, apart from short-term adjustment problems, or problems arising out of longer-term structural rigidities in the economy, which should be tackled by appropriate measures to increase the flexibility of the economy, there is no fundamental choice to be made between jobs and pollution abatement. In so far as policies lead to a reduction in pollution they imply a *shift* in the way the economy uses its resources, not a change in the *total* amount of resources used. If, for a given use of resources, total final output changes in response to a reduction in pollution, the authorities may have a minor problem of controlling the pressure of demand in the economy. But governments are constantly concerned with this problem anyway, since, in addition to continuous changes in the pattern of demand and output, the variables determining the total pressure of demand are constantly changing. In any event, the total macro-economic burden of environmental protection is probably negligible.

QUESTIONS FOR DISCUSSION

1. 'Pollution destroys the environment. It should therefore be stopped at all costs'. Discuss.

2. 'Economic growth is the cause of pollution'. Argue for and against.

3. 'The environment is polluted because no-one owns it.' Do you agree? Illustrate your answer by parts of the environment that are and by parts that are not owned.

4. What is the economic significance of the scarcity of pure air, clean water and other constituents of the environment?

5. An objection from industry to charging is that it would raise costs. Discuss.

6. Would charging be less or more expensive than control by direct state regulation?

7. Pollution charges would advantage large firms because they could (a) pay them, (b) pass them on, more easily than smaller firms. Discuss.

8. What is the economic inter-relationship between 'optimum' pollution and the distribution of income?

9. Discuss five main objections to pollution charging. How do they compare and contrast with the objections to direct regulation?

10. Discuss the impact and the incidence of pollution charges on industry. To whom should the proceeds be paid? Why?

FURTHER READING

BOOKS AND ARTICLES

Baumol, W. J., *On International Problems of the Environment*, Wicksell Lectures 1972, Almqvist and Wicksell, Stockholm, 1972.
'On Taxation and the Control of Externalities', *American Economic Review*, June 1972.
and Oates, Wallace, *The Theory of Environmental Policy*, Prentice-Hall, New Jersey, 1975.

Dales, J. H., *Pollution, Property and Prices: An Essay in Policy-making and Economics*, University of Toronto Press, Toronto, 1968.

Dorfman, Robert and Nancy, *Economics of the Environment: Selected Readings*, Norton, New York, 1972.

Edel, Matthew, *Economics and the Environment*, Prentice-Hall, New Jersey, 1973.

Jacoby, Neil H., and Pennance, F. G., *The Polluters: Industry or Government?* Occasional Paper 36, Institute of Economic Affairs, 1972.

Kneese, A. V., and Bower, Blair, *Managing Water Quality: Economics, Technology, Institutions*, Johns Hopkins Press for Resources for the Future, Baltimore, 1968.

Meade, James E., *The Theory of Economic Externalities*, Sijhthoff, Leiden, 1973.

Pearce, D., Markandya, A., and Barbier, E. B., *Blueprint for a Green Economy*, Ch. 7: 'Prices and Incentives for Environmental Improvement', London: Earthscan, 1989.

Pezzey, J., 'Market Mechanisms of Pollution Control: "Polluter Pays", Economic and Practical Aspects', in R. K. Turner (ed.), *Sustainable Environmental Management: Principles and Practice*, London: Belhaven Press, 1988.

Swedish Journal of Economics: Special Issue on Environmental Economics, March 1971.

Wirth, T., and Heinz, J., *Project '88/Harnessing Market Forces to Protect our Environment: Initiatives for the New President*, Washington DC, 1988.

OFFICIAL REPORTS

OECD, *The Polluter Pays Principle*, Paris, 1975.

OECD, *Pollution Charges in Practice*, Paris, 1980.

OECD, *Economic Instruments for Environmental Protection*, Paris, 1989.

Royal Commission on Environmental Pollution:
First Report, Cmnd. 4585, HMSO, London, February 1971, Chapter 2.
Third Report: Pollution in some British Estuaries and Coastal Waters, Cmnd. 5054, HMSO, London, September 1972, especially, Minority Report by Lord Zuckerman and Dr Wilfred Beckerman.